Ecoviews Too

Ecoviews Too

Ecology for
All Seasons

Whit Gibbons
and
Anne R. Gibbons

THE UNIVERSITY OF ALABAMA PRESS
Tuscaloosa

The University of Alabama Press
Tuscaloosa, Alabama 35487-0380
uapress.ua.edu

Inquiries about reproducing material from this work should be addressed
to the University of Alabama Press.

Set in Palatino TT and Open Sans types
Manufactured in the United States of America
Cover image: Dover Pictorial Archive
Cover design: Gary Gore

Cataloging-in-Publication data is available from the Library of Congress.
ISBN: 978-0-8173-5875-4
E-ISBN: 978-0-8173-9083-9

For my grandchildren,
Allison, Parker, Sam, and Nick, and their grandmother
Carolyn; may you never lose your sense of wonder and
your enthusiasm for nature's marvels.

—WHIT GIBBONS

For my husband, Bill Fitts,
with thanks for his never-ending love and support.

—ANNE R. GIBBONS

Authors' Note

This book was written by both of us, a collaboration reflected in the first person plural pronouns used in the introductory and concluding material. For the sake of clarity and consistency, however, we have chosen to retain the use of the first person singular as it appeared in the original newspaper columns.

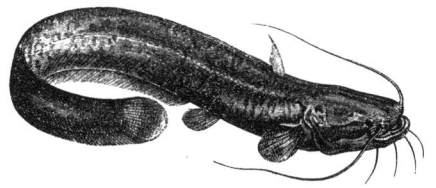

Contents

Autumn

Winter

Acknowledgments

Producing a book is always a collaborative effort. In the case of *Ecoviews Too*, thanks are due to the myriad people—scientists, researchers, and laypeople; colleagues, friends, and family members—who read the newspaper columns prior to publication: Kimberly Andrews, Steve Bennett, Kurt Buhlmann, Vince Burke, Mike Dorcas, Luke Fedewa, Peggy Fitch, Xavier Glaudas, Gabrielle Graeter, Aaliyah Green, Judy Greene, Susan Harris, Meg Hoyle, Tom Hughes, Linda Lee, Yale Leiden, Tom Luhring, Ken McLeod, Brian Metts, Mark Mills, Tony Mills, Joe Pechmann, Brian Todd, Ria Tsaliagos, Tracey Tuberville, Phil Vogrinc, Lucas Wilkinson, J. D. Willson, and Chris Winne. We extend special thanks to Teresa Carroll, Sarah Collie, Margaret Wead, and Patricia West for their many years of assistance in reviewing columns and gathering information.

Several cogent recommendations made by two of the University of Alabama Press's anonymous manuscript reviewers prompted us to ask permission to formally acknowledge their contributions. We therefore thank Robert W. Hastings and Ken R. Marion for their constructive comments and suggestions. Thanks also go to Beth Motherwell at the Press for encouraging us to undertake the project as a sequel to our first book, *Ecoviews: Snakes, Snails, and Environmental Tales* (University of Alabama Press, 1998) and to Susan Harris for her careful copyediting and indexing.

For participating in outdoor adventures on which

some columns were based we thank Mike, Jennifer, Allison, and Parker Gibbons; JoLee Gibbons Passerini; Keith, Susan, and Nick Harris; Laura Gibbons and Ron Curtis; Jennifer, Jim, and Sam High. Finally, we are especially appreciative of Carolyn Gibbons and Bill Fitts for their support and encouragement during the book-writing process.

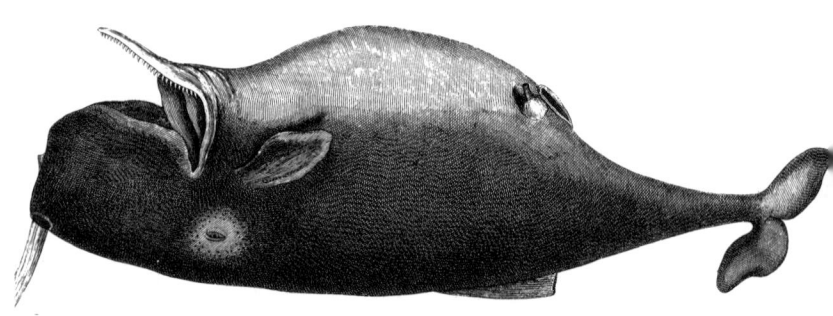

Preface

Nature, in all its myriad and amazing manifestations, can be enjoyed in every season. Sometimes the predictability of a natural event is what appeals. Other events are treasured for their rarity or unpredictability. And sometimes—such as when rain changes from life giving to life destroying—nature wields a double-edged sword.

Referring to a plant's or animal's "environment" is commonplace, implying that living organisms dwell in and respond to a set of constant, unchanging conditions. But with few exceptions, such as caves and the ocean's darkest depths, every place on Earth experiences continual and often dramatic temporal changes. The transformation from day to night is ubiquitous and predictable. In the temperate zones the environment is constantly changing because of the seasons, with unpredictability being the most predictable feature. Cold winters, except for unusually warm ones, and hot summers, except for greatly appreciated cool ones, are expected. Extended droughts punctuated by periodic floods are the norm over a period of decades. In some tropical regions the year is partitioned into wet and dry seasons, with the extremes varying annually.

A new year heralds a new beginning. But deciding when the New Year begins depends on who is doing the deciding. January 1 opens the door for standard calendars, whereas the US government fiscal calendar begins on October 1. For many universities and colleges, the New Year begins on July 1, to accord somewhat with the academic year. And throughout history, various cultures have defined the beginning and end of a year by dates that typically related to seasonal celestial changes. For the plants and animals of the natural world, various activities associated with life cycles—courtship, mating, birth and other reproductive activities, growth, dispersal, and countless other vital functions of being a particular species—determine when the New Year begins.

For our purposes in this book, the New Year will commence in "springtime" as perceived by someone living in temperate North America in the southern United States. Springtime is when it's not necessary to heat the house on a daily basis but air conditioning has not yet been turned on. The few weeks of springtime, and its autumnal counterpart, are cherished by most people in the South. This is also when wild plants and animals begin to take ownership of their outdoor habitats in a manner appropriate for their own biological trajectories.

This book, an environmental anthology, is based on the weekly column Ecoviews that has been published by the *Tuscaloosa (AL) News* and other newspapers for more than thirty years. The four sections of this book, represented by the seasons of the year, track the adaptations and responses of wildlife to the inexorable changes that occur at any location over the course of a year. The ecological vignettes focus on seasonal happenings, particular holidays, or, in some instances, historic events that define a moment when the connection between society and ecology was fundamentally altered.

Introduction

What Should You Do with a Beached Whale?

Imagine walking out to get your newspaper one morning and finding a live whale stranded in shallow water near the porch. For this to happen, you would of course have to live on the coast, with water close by your front door. Assuming those things to be true, how would you feel about the stranded whale? The first response would surely be wonderment. (If you would not be at least temporarily surprised, then you have led a significantly different life from the rest of us.) But after a moment of disbelief, how would you feel?

Over the past forty years wildlife biologists have addressed and attempted to categorize the disparate views individuals take toward encounters with wild animals. The eleven perspectives listed below are our interpretation of the fundamental attitudes. These divergent views about wildlife are the underlying cause of many environmental conflicts because no single sentiment is necessarily right to the exclusion of the others. A further complication arises because most people's responses include a combination of two or more attitudes. But virtually no one is going to express all eleven or even the majority of the attitudes about a single situation because some are in direct conflict with each other. However, the same individual might respond differently at different times and in different situations.

Following is our list of eleven classic attitudes humans have toward wildlife:

1. Compassionate—characterized by a nurturing response toward wildlife. Such a person would seek help to move the beast back into the ocean to save its life. This animal welfare approach conflicts with some other environmental attitudes.

2. Scientific—characterized by a desire to study and understand wildlife. A scientist is objective and impartial. What species of whale is it? Why is it stranded? Does it have an inner ear parasite that caused disorientation? Perhaps we should let it die so we can dissect it. A research ecologist fits easily into this category.

3. Environmental—characterized by the view that an individual animal is of less significance than the species. Some ecologists might think if a whale is dumb enough to beach itself, it will be better for the species if its genes are not passed on. Concern would focus on whether something is causing a die-off of whales in general.

4. Aesthetic—characterized by an interest in the symbolic or artistic aspects of animals. Having such a magnificent animal in your yard would be a pleasurable experience. A few photographs or an oil painting could be appropriate. On its demise, a poem might be in order.

5. Utilitarian—characterized by an interest in the practical value of the animal. What good is it to me and how can I take advantage of this newfound commodity? Can I charge scientists to study it or artists to depict it? On a broader scale, could some aspect of a whale's anatomy lead to a cure for cancer?

6. Dominating—characterized by a desire to pursue wildlife for sport, sustenance, or both. The hunting instinct could emerge if the whale began escaping into deeper waters. A hunter might want to herd it back, capture it, and then possibly let it go. If no regulatory con-

sequences were likely to be forthcoming, some hunters would kill the animal.

7. Negative—characterized by fear or dislike of animals. For almost everyone, wildlife can be bothersome in certain situations (think blood-sucking mosquitoes and garbage-strewing raccoons). But some people avoid the outdoors because they find wildlife consistently annoying or threatening. For such people, a fifty-ton whale would almost certainly be considered a nuisance animal that should be removed by the wildlife department.

8. Managerial—characterized by a desire to manipulate and control all aspects of one's environment. The consummate manager would probably want to take measures to ensure that whales could no longer do this stranding thing to themselves because no one is in charge and it annoys coastal residents.

9. Indifference—characterized by a lack of interest in wildlife. Sadly, this attitude is not uncommon, although it would seem an unlikely response to finding a stranded whale in one's yard.

10. Naturalistic—characterized by an enthusiasm for wildlife and the outdoors. A beached whale would obviously invoke wonder, but pleasure would also be found at the sight of shorebirds and woodland foxes, wildflowers on dunes or along mountain paths, and fireflies in one's own backyard.

11. Religious—characterized by a belief in a higher power that is connected to everything in nature. Some people might seek to deal with the situation through prayer; others might ascribe the whale's plight to predestination. Some might deem a whale in the yard an omen worthy of further consideration.

Obviously, a stranded whale isn't an everyday ocurrence, even if you live near the ocean. But the environmental attitudes described above are applicable to even the most mundane encounters with wildlife.

The environment has now joined politics and religion as a controversial and potentially divisive topic of conversation. Accepting the views of others is not always easy, particularly when those views are directly opposed to our own—but it's something to strive for.

Instances of cooperation to further the common good have produced some of our nation's greatest political environmental achievements: creation of the Environmental Protection Agency and passage of the Endangered Species, Clean Air, and Clean Water Acts, as well as legislation to protect wetlands. Targeted attacks to weaken environmental laws and regulations are generally ill founded, with rewards benefiting the special interests of only a few individuals or companies. Most federal environmental programs need strengthening, and none of us should have to endure concerted efforts to weaken or eliminate environmental safeguards. The broad intent of political efforts to protect natural habitats is in the best interest of the human population of the United States and the world.

Our goal for this book is to encourage, maybe even inspire, others to appreciate natural habitats and the native wildlife that populate them.

Spring

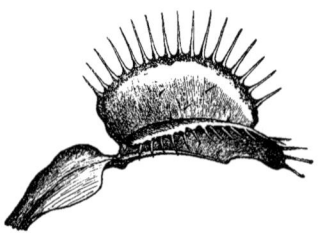

Every Day Is April Fool's for Some Species

A fish spies a wiggling worm under the riverbank. Free meal? Yes, but not for the fish. April Fool's. The would-be worm was actually the tongue of an alligator snapping turtle, and the giant jaws slammed shut when the fish went after the bait. For almost any identifiable human behavior, including playing jokes on April 1, an equivalent or near-equivalent can be found somewhere in the plant and animal kingdoms. For many species, playing tricks is a daily routine. But doing so is definitely no joke; it's a matter of life or death.

Successful predators that rely on capturing living prey must have effective strategies for doing so, and subterfuge is one such technique. A strategy as basic as a bobcat's use of camouflage while stalking a rabbit in a field of brown vegetation is clearly meant to deceive and catch the unwitting prey off guard. Likewise, a tiny deer fawn's response to potential danger is to lie motionless on the forest floor, where the white spots on its brown coat blend with the ground speckling of sunlight through the leaves, perfect camouflage to trick a would-be predator.

Luring prey to its death is a scam used by many predators. Baby copperheads, cottonmouths, and pigmy rattlesnakes wiggle their bright yellow tails to attract small frogs or lizards. Some tropical lightning bugs flash the mating signal of other species to attract males that think they are headed toward a romantic encounter only to become a meal for the deceiver. Lights are also used as lures by deep-sea angler fish. These big-mouthed creatures, which live in virtually total darkness at ocean depths greater than a mile, have a fleshy structure that functions like a fishing pole with bait at the end. The lure is a luminous bulb containing bacteria that emit a greenish light to attract other fish, which are fooled into thinking they will be getting a meal instead of becoming one.

Some plants rival animals in their use of chicanery to capture prey. Well-known carnivorous plants include pitcher plants with their sweet-smelling but deadly pitfall traps. Plants that eat animals typically live in highly acidic wetland habitats that are low in soil nutrients. Their captured prey, mostly insects and spiders, provide some needed nutrients. Among carnivorous plants, the showy Venus flytrap has an impressive April Fool's surprise for visiting insects. The two halves of an open flytrap leaf look innocuous enough but have long spinelike structures extending out from the edges. Nectar glands on the inside of the leaf signal a tasty meal for flies and other insects. When a bug alights and its legs begin to hit hair triggers, the trap slams shut so fast even a fly cannot escape. Over the next several hours the flytrap secretes digestive juices that absorb the prey.

To eat or to avoid being eaten are not the only reasons to engage in trickery. Strategies used to acquire mates involve some of the most duplicitous behavior found in nature. A predaceous insect known as the scorpionfly definitely ranks high on the deceit-o-meter for its mate-luring behavior. Male scorpionflies impress females by presenting them with a blowfly acquired at great personal risk from a spider's web. The female's acceptance of the blowfly dinner assures the male of a mating opportunity. But some male scorpionflies do not capture their own blowflies. Instead they pose as females in order to fool another male scorpionfly into handing over his hard-earned gift. Once the male with a blowfly offers his tasty treat to the male poseur, the deceiver accepts the gift and flies away to use the pilfered blowfly to attract a female for mating.

From simple to complex, the diverse tactics used by wildlife to get food, protect themselves, or acquire a mate provide endless reasons to marvel at the natural world. For these deceivers, life or death matters hang in the balance and they do not wait for the first day of April to play their tricks.

Tips for Earth Day and Proper Environmental Etiquette

I have two environmental suggestions for Earth Day. As almost everyone who likes to breathe clean air knows, Earth Day falls on April 22. But one day is not enough. We should celebrate Earth Day year-round, from the time we wake up until we go to sleep, all day, every day. After all, we want a clean, healthy, and enjoyable environment all the time, not just one day a year. So my first suggestion is that we start celebrating Earth Year and make Earth Day something we did last century.

Many people already have the Earth Year attitude and contribute to making our environment safe, healthy, and pleasant in myriad ways. One of my favorite activities, which I hope is a contribution, is to answer questions that help people become familiar with the intriguing array of life forms and ecological interactions that exist in the world. The complexity of both exotic and everyday plants and animals is fascinating, and I applaud everyone who wants to learn more about the natural world. So my contribution to Earth Year is to answer questions and respond to opinions related to environmental awareness and appreciation. Which brings me to my second suggestion: people who have environmental queries should exercise at least minimal email etiquette.

I am not able to check my email account every day but I do try to respond to all queries eventually. Of course, if your question is like this one—"I have a copperhead coiled up on my front porch. Should I let the dog and children play in the yard?"—my answer may be too late to help you. Sorry. But the timeliness of my responses is not the issue. My point is that the basic elements of any email from a polite and reasonably intelligent person should include who you are and where you are.

I like to be responsive to the people who write me. But how much time should I invest in an email that says, in its entirety, "I would like to know how to raise catfish. Please respond as soon as possible." I did send the anonymous writer from an anonymous place the link to a website on catfish. Should I have done even that? Another one that is more frequent than I care to remember runs something like this: "I am doing a report for class and would like for you to send everything you can about ecology and the environment. My report is due tomorrow. Sparky." These are the kinds of emails I like to open a day late.

And how should I respond to "I saw a black snake crossing the street. What is it?" No herpetologist could give a definitive answer without knowing—at the very least—what continent you saw the snake on. The list of problematic emails I have received goes on and on. But it's time to bring this rant to a close and offer four simple rules for composing an email in which you are asking someone for information or advice.

1. Give your name and, if appropriate, your affiliation with a school or organization.

2. Indicate where you live. Remember that with most

Q: What is a normal day like in the field for an ecologist?

A: There is no "normal" day in ecology. Environments change continually, so each day can present a new adventure. In field ecology programs, most ecologists are impressed at how many new discoveries continue to be made about what animals and plants do.

emails someone only knows what planet you live on. An important clue for identifying some plants and animals is specific location. If you are requesting information about a topic for an academic project, identify the course you are taking and where you go to school.

3. Plan for the possibility that you may have to wait a few days for a response.

4. Write your message in English, not text and online chat abbreviations. And take time to proofread it before you press Send.

So, those are my two suggestions: First, one day is not enough to honor Earth and all its glories. Let's celebrate Earth Year, each day, all day long. Second, I commend everyone who wants to learn about animals, plants, and ecology. Just remember to follow the simple rules of email etiquette outlined above to ensure that you get a helpful answer in a timely fashion.

Easter Is Associated with Many Plants

Easter sends various messages to people throughout the world. Of particular interest from an ecological standpoint are the many trees and flowers that are associated with that time of year. Flowering dogwoods, redbuds, palm trees, lilies, and many other plants have connections with Easter, some well known, others less so.

One of the best-known stories involves the flowering dogwood tree. According to legend, dogwoods once grew to be the size of oaks and were used to make the cross on which Jesus was crucified. One version claims that Jesus saw the dogwood as suffering because of its having been used for such a purpose and avowed that the tree would never again grow to a size that would allow it to be used to make a cross, hence the small, crooked branches. Other symbolic features of the dogwood tree are white "flowers" that form the shape of a cross, with a brownish red spot in the center of each, signifying Jesus's blood. The cluster of tiny flowers in the center resembles a crown of thorns.

In the preceding paragraph, "flowers" is in quotes because, as botany students are fond of pointing out, the true flowers of a dogwood are the tiny ones in the middle. The more obvious "petals" that have earned the tree the name "flowering dogwood" are actually bracts. In most plants, bracts are small leaves from which the flowers arise. In the dogwood they have taken on a more prominent appearance. Incidentally, the popular dogwood familiar to everyone from New England to Florida is one of ten species in this country. Ironically, as far as I am aware, dogwood is not one of the more than eighty kinds of plants mentioned in the Bible.

One type of tree mentioned prominently in many parts of the Old and New Testaments is the palm. Preceding Easter is Palm Sunday, which refers to the fronds of date palms that were placed in front of Jesus's donkey when he entered Jerusalem, as was done for a victorious ruler. I do not know of any other special significance to palms, other than that some churches pass palm fronds out to members of the congregation on Palm Sunday. Although native to the Mediterranean region, date palms made their way to Mexico via Spanish explorers and eventually arrived in California in the 1700s. Today more than a quarter of a million date palm trees bear fruit in California and Arizona. Like many trees, date palms are unisexual, which means an individual tree is either male or female. Since only the females bear fruit, people who plant palm trees as a commercial product want mostly females, with only a few males for fertilization. The sex of a tree can be assured by planting small shoots that grow from the base of the palm trees.

The crown of thorns placed on the head of Jesus is assumed to have been a particular type of shrub, a member of the rose family called thorny burnet. The plant is and was abundant around Jerusalem and other parts of the Mediterranean. The wooden branches are flexible enough to bend, and the thorns at the end also branch. Other types of plants with briars and spines are found in the region, but the thorny burnet is most likely the one that was used for the crown of thorns.

One plant associated with Easter is purely a commercial venture, albeit an agreeable one. The original species of Easter lily is native to the Ryukyu Islands, halfway between Japan and Taiwan, and had nothing to do with the Middle East. Today, the vast majority of Easter lilies come from agricultural lands from Oregon to California and show up in churches across the nation. They may be symbolic of the season in people's minds, but I know of no biblical connection.

Finally, another tree not mentioned in the Bible is the redbud tree, also called the Judas tree. The Mediterranean species of redbud tree is said to be the one on which Judas Iscariot hanged himself. According to legend, redbud trees turn red in the spring, either from blushing for shame at the crucifixion of Christ or from weeping tears of blood at the fate of Judas.

Mothers of Many Animals Are Worthy of Recognition

Cowbirds, starfish, and turtles do not give or receive Mother's Day gifts. One reason is that offspring of these animals do not know their mothers. Cowbirds deposit their eggs in the nests of other birds, and unknowing foster parents raise baby cowbirds along with their own young. Turtles lay their eggs in dirt or sand and never look back. And a detached starfish arm can grow into a complete starfish that cannot claim to have had even a neglectful mother.

Nonetheless, humans do not have a monopoly on maternal devotion. The ancestors of many animal species here on Earth today successfully jumped the high hurdles of evolution simply because they had good mothers. Although young alligators do not bring their mothers flowers or candy, alligators join humans and many other mammals in representing the kind of maternal care that warrants recognition on Mother's Day. These mothers are attentive to their offspring before birth and long after. All will do what they can to protect their eggs or babies from harm. At the other extreme are the mothers of most amphibians, reptiles, insects, and fish, which lay eggs in carefully chosen spots but then disappear. The eggs and young are on their own for the rest of their lives.

But even these species still deserve a Mother's Day card for their front-end investment in their offspring. For example, a female slider turtle develops large follicles, which are equivalent to the yolk of a bird egg, months before laying eggs. When the eggs are fertilized by a male slider during the spring mating period, the developing embryos have enough yolk to nourish the baby turtle for an entire year. The mother lays the eggs in what she intends as a safe, underground nest. So, although she drops the eggs, covers the nest, and never looks back, she has done her motherly duty before the baby ever reaches the water.

And exceptions exist even among these groups. Burmese pythons, the giant constrictors that have become established in southern Florida, are noted for unusual maternal behavior. The female not only coils her body around the eggs but also warms herself up by shivering, thus enhancing the incubation of the eggs. Female king cobras, the largest venomous snakes in the world, reaching lengths of eighteen feet, are reported to stand guard until their eggs are safely hatched. A predator considering eating the eggs of either pythons or cobras might well become a meal itself. The newborn young of diamondback rattlesnakes have been reported to remain with their mothers for up to two weeks, surefire protection from most predators.

Mother blue-tailed skinks, common lizards in the eastern United States, stay with their clutch of eggs until they hatch. If a baby dies in the egg before hatching, the dutiful mother will pick the egg up in her mouth and remove it from the nest. In the social insects such as wasps and ants, the entire colony works to protect the young.

Certain species of frogs are contenders for the best-mother award among cold-blooded animals. Full-grown adult female Jamaican frogs are tiny, less than two inches in length. The mother lays her eggs, about fifty of them, up to 250 feet deep inside a cave and stays with the developing eggs. After about a month they hatch, and the baby frogs crawl up on their mother's back for the trip out of the cave. The mother can jump more than three feet without losing any of the babies! Hopping through a dark cave with babies clinging to your back is a clear demonstration of maternal devotion.

Most female birds exhibit parental care by at least incubating the eggs. In some cases both parents provide care even after the babies hatch, but the mother gets credit for laying the eggs and always being around till the young are ready to fledge. Anyone who spends much time watching backyard birds around bird feeders has seen female cardinals or house finches feed sunflower seeds to obvious juveniles. The baby, about equivalent in age to a human teenager, flies over with its mouth open and wings aflutter, still looking for a handout, which the mother provides. Readers may draw their own parallels with the behavior of human teenagers.

Like humans, all mammal babies depend on mother's milk for nourishment. Even so, the variability among mammal species is great. Two mammal species, the duckbill platypus and spiny anteater of Australia, lay eggs. Tree shrews are mammals that give minimal attention to their young. The female lives with her mate in one tree and has her babies in a nest some distance away. She visits the nest once every two days to let the young nurse. Like other mammals, mother whales, porpoises, and manatees nurse their young, but they also have a unique role to play in rearing their young: they nudge them to the surface at regular intervals for air. Meanwhile, the marsupials, such as kangaroos and possums, not only nurse their babies but also carry them around in a pouch until they can fend for themselves.

Can we declare which animals make the best mothers? No. The parents of every species do what works best for them based on their evolutionary history. Any species that is still around has presumably been doing things right, whether by constant attention or benign

Q: We sometimes see raccoons in our neighborhood, usually at night. Recently, our dog chased a large one across the yard during the daytime. The raccoon got away, and the dog was nosing at a very small baby raccoon the mother had dropped. I got the dog inside the house. Later that day, the baby was gone. Would the mother have come back for it?

A: The mother probably came back for the baby, which she would then have carried to safety. Raccoons will not ordinarily protect their babies out in the open, the way bears would, but they will move very young babies from one place to another if they feel threatened.

neglect. The measures animals will take to protect their offspring are more impressive for some than for others. But the mothers of all animals, including humans, are so exceptional that it is fitting to have a special day to honor them.

Some Birds Take Care of Their Siblings

Spring is the time to enjoy nature's productivity. As we approach the official first day of spring each year (around March 21), we can hear more birds singing, see pairs flitting here and there, and watch nest-building activities. Wrens started a nest in our garage this week, which means we have to leave the doors open till the young leave. It's worth it to be able to watch the birds and their fledglings.

But one aspect of nesting birds is less charming and might send chills through a typical bird-watcher: siblicide, the killing of a brother or sister. Siblicide in the nest has been documented for several bird species. Siblicidal birds are not cannibalistic. They do not eat their siblings; they just kill them.

Apparently, siblicide is the order of the day for some birds. For example, the hatchlings of black eagles of Africa and the Middle East seem to be a species on a death mission. Black eagles lay two eggs. Yet of two hundred nests examined by a research team, only one contained two young that survived to fledge. The deaths resulted from direct attack by the victim's sibling. In one nest, the larger eaglet pecked the smaller one 1,569 times before death occurred. In no instance did the parent birds become involved in fights between siblings; maybe they assumed that baby birds will be baby birds.

Black eagles are not the only birds that engage in such behavior. According to studies, siblicide is characteristic of many other birds, including pelicans, ospreys, and other eagles that lay two eggs. Some egrets are also notorious killers of their siblings in the nest. One scientific publication documented that a pair of young cattle egrets pecked a third one senseless. The helpless victim, evicted from the nest, fell to the ground and died.

In searching for similarities among bird species in which siblicide occurs, it comes as no surprise that birds that kill one another in the nest possess the weapons to do so. Large, pointed or hooked beaks aid in pecking one's nestmate to death. Also, such species are confined to an enclosed nest area, in contrast to many other birds, such as killdeer and bobwhite quail that lay their eggs on the ground. In these species, the young scatter soon after they hatch. Perhaps they heard cautionary tales about what happens to little birds that hang around the nest too long.

Another aspect of siblicide relates to food resources. Siblicidal species are usually ones for which an adequate supply of baby bird food cannot always be assured. This is not the result of parental neglect; it is caused instead by a situation, such as drought, in which parents cannot find the constant, abundant supply of food necessary to keep all their young healthy and happy. This sets up a competitive relationship for the babies.

A trait identified as very important in siblicidal behavior is that some disparity in size exists between the nestmates. For example, in the black eagle, the mother hatches the two eggs three days apart, assuring that one will be larger than the other. Then, if food supplies are low in a particular year, the smaller nestling is killed by the other. If the first nestling is unhealthy in some way, the second one becomes dominant. Either way, the parents can count on at least one well-fed offspring.

In seasons when food is plentiful, more than one

Q: I have just realized that I no longer see thrashers, blue jays, or the once-plentiful woodpeckers in our yard. Could this have anything to do with the loss of chipmunks due to the rat poison we put out after we saw rats sitting on our air conditioner enjoying the nearby bird seed? Or is it due to several cats that roam the neighborhood? Or is it a more widespread problem?

A: Rat poison and cats can certainly be indicted for some changes in species composition that might be observed in a backyard, but change in the abundance of certain animal species at a given location over time can also be a natural phenomenon. Almost everywhere, some species are more abundant than in previous years whereas other species seem to be on the decline or even absent entirely. These so-called wildlife cycles are not just perceptions, they are real. And what causes them has puzzled and intrigued fur trappers, farmers, and ecologists for well over a century.

nestling may be fed successfully and siblicide may not occur. From the human perspective, siblicide as a means to ensure survival of the fittest may seem severe. But the phenomenon is widespread in the bird world and is clearly a successful strategy.

Birds also have other methods for dealing with limited food resources. Many birds, including red-cockaded woodpeckers, have evolved a system of "helpers," usually immature birds from an earlier nesting, that assist the parents in the care and feeding of young. This is not to suggest that birds are basically kind to one another; the helpers acquire experience in nesting, an advantage once they are of an age to become parents themselves.

This spring, if you see a baby bird that's fallen from the nest, it may not be an accident but a decisive act by a sibling, maybe even a parent in some instances. Don't pass judgment on such behavior. The actions of any animal species, whether they seem cooperative or cruel to us, are generally those behaviors with the highest probability of passing the parents' genes on to the next generation. Nature's ways are exceedingly complex.

Bumblebees Can Make Honey, Too

I received a question from someone who wanted to get information about bumblebees, these creatures that appear in our backyards as a buzzing cloud of yellow and black. The writer said that each summer day hundreds of them blatantly pilfer the pollen from the many flowers in her yard. She was familiar with hummingbirds and butterflies, but her knowledge concerning the bumblebee was woefully lacking. She had several specific questions: Where do they nest? Is it a hive? Do they make honey? What is the purpose of their pollen gathering? Does the normal bumblebee mate or only the queen? Do they sting? Do they have a natural enemy?

Bumblebees, with their rounded, robust bodies, remind me of linebackers that play for a team whose school colors are black and yellow. They are highly watchable creatures to have around a yard. Most bumblebees, of which there are more than two hundred species worldwide and fifty in North America, are social insects that live in a nest they build, often in an abandoned mouse nest. I have a bumblebee nest in my yard that is in a bird box that had been the home of flying squirrels this past winter and spring. Like their close relatives the honeybees, bumblebee colonies have a queen. The queen is recognizably larger than the worker bees, all of which are female. Apparently queen bumblebees will sometimes leave the nest to forage like the worker bees. Male bumblebees are known as drones, and their sole purpose is for mating with the queen.

Like some of the other social bees and wasps, bumblebees will sting, but only if threatened or provoked,

Q: What is the strange-looking creature I see in a slow-moving stream in South Carolina that resembles a spider that can walk across the water? Sometimes several are scooting around on the water's surface. Do they bite?

A: That is one of the water striders. More than seventeen hundred different kinds are found worldwide. A water strider looks superficially like a spider skating across the water surface. From above you can see where the pair of front and hind legs create little indentations from water tension on the surface. It looks like it has pontoons for feet. The middle legs are used like a pair of oars for rowing across the water. Their movement seems effortless as they glide smoothly and rapidly across still water. Studies conducted on water striders have determined that their primary form of communication with each other is through leg vibrations that send ripples across the water. One message might tell a nearby water strider to keep his distance, whereas other vibrations from a male can tell a female to come hither.

I have never heard of a water strider biting a person, but I once saw a tiny green caterpillar wriggling in a slow-moving stream. Upon closer inspection I saw that it was held by (and presumably being devoured by) a water strider. Four other striders were moving around it but keeping their distance. Presumably a ripple message from the one with the caterpillar was saying, "Mine. Mine. Stay away." A lot of action is in progress in the natural world around us, even among some of our smallest species. Take a stroll and see what you can find.

such as when the nest is disturbed. Drones have no stingers and they do not gather pollen like the workers. Most bumblebees are extremely benign, especially compared to yellow jackets or hornets, which protect their nests with a vengeance. I have never heard of a bumblebee stinging anyone who simply stood and watched one gather pollen on flowers, although they might sting anyone that disturbs a nest and stays to have a closer look. Catching one by hand would of course be unadvisable.

Honeybees and bumblebees are attracted to flowers to get the sweet-smelling nectar and pollen. During the nectar-gathering process, insects such as bees also transfer pollen from one flower to another, serving an essential role in pollination, a requirement for some plants to successfully produce fruits and seeds. Nectar is returned to the nest by the bee, allowed to dry to a certain extent, and then in a complex process that involves bee saliva and special enzymes, is eventually converted to honey. Honey is used to feed members of the colony. Bumblebees produce honey but never in the quantities found in honeybee hives. Bears probably spend little time looking for bumblebee nests to raid.

Even in the southern United States, most bumblebees do not last through the winter, except for young queens that emerge from the nest and mate with drones from other colonies. A future colony queen finds a safe place, such as underground or in a tree hole, to hibernate through the winter. In the spring she seeks out a site for the nest, where she lays eggs to begin a new colony. The queen can control the sex of her young, and early in the season she produces only infertile females, the workers. The queen gathers nectar and pollen to feed the first of the developing larvae and pupae, which begin their role as worker bees that take care of the nest and the young that are produced later.

As far as natural enemies, adult bumblebees probably have few because most predators would prefer not to risk being stung for such a small meal. Ironically, one of the greatest threats to bumblebee colonies in some areas is another kind of bumblebee, marauders known as nest parasites that sometimes kill the queen of the colony, lay eggs, and have their own developing young fed by the bumblebees of the colony. We do not have to go far from home to be surrounded by ecological mysteries and intrigue in the everyday wildlife around us.

We Don't Need to Kill Carpenter Bees

Every year at this time I get questions about bees. Recently, a friend told me about honeybees swarming on a university campus. She and her fellow employees gathered to watch as the bees flew around the office windows then collected in a huge clump on a nearby tree. The bees were still there the next day when my friend got to work, but twenty minutes later they had disappeared. Not a bee was to be seen. What was going on?

A queen and several thousand workers were simply looking for a new home. The bees hang in a large cluster while scouts look for a place to start a new hive. Honeybees typically do not sting during the swarming stage. My son and I once put a cardboard box with a hole in it next to a swarm and let the bees crawl over our hands as they walked inside to set up a new colony. I was pleased to find that what I had been told about bee swarms not stinging was true. We took the box of bees home, put them in a wooden box, and a couple of years later actually ate honey they produced.

Honeybees have declined across the United States since the 1990s. Explanations for the decline include infestations of parasitic mites in the nests and the use of pesticides. A swarm of bees indicates that a nearby hive has become overcrowded and some bees have left. A healthy sign.

Another bee question concerns tiny piles of sawdust from unpainted wooden structures. The sawdust is caused by carpenter bees that burrow into the wood. Carpenter bees are similar to bumblebees in being about an inch long but without the fuzzy yellow appearance of bumblebees. Carpenter bees can sometimes sport a two-tone look when carrying a supply of yellow pollen. A common question is, how can I get rid of them? The simplest and best answer is do not get rid of them. Enjoy them. On our back porch, instead of eradicating them, we watch them, listen to them, and smile as the dog snaps at them. Cheap entertainment on a spring day.

The ecology of carpenter bees is relatively straightforward. In spring, throughout most of the country, they seek mates. Males, which have a distinctive light-colored spot on the face, may buzz loudly and fly in front of a person's face. But the bluster is just an act. Male carpenter bees are harmless. As with other bees and wasps, only the females have stingers. Some entomologists consider the in-your-face behavior to be curiosity instead of aggression. My grammar-school-age grandsons now make a game of air-grabbing carpenter bees, making sure the yellow spot on the head is visible to show it is a harmless male. The bees fly away when released, but the same ones will eventually come back. They return with a more wary attitude toward little boys.

Female carpenter bees literally chew a tunnel into wood. I know they work at night as well as day because I have watched a steady stream of sawdust trickle from the ceiling an hour after dark. The females prefer an already created hole, but some additional excavation and reorganization may be part of the process, much like someone moving into a new apartment might paint the walls or rearrange the furniture. Sitting beneath a carpenter bee reconstruction project leads to frequent brushing of sawdust from your clothes. Having a visitor sit in the chair beneath a carpenter bee reconstruction project can be entertaining—at least to the host.

Carpenter bees are valuable pollinators—females gather pollen, store it in the burrows, and lay their eggs. The pollen serves as a source of nutrition for the larvae.

The adult bees die during the summer, and the recently born ones spend the winter in the holes. Certain people's response to carpenter bees confirms that the control mentality is sometimes unnecessary and unreasonable. Many cooperative extension units associated with universities provide information on how to control carpenter bees with pesticides. Come on! Haven't we learned by now that pesticides never kill just the target organism but many other harmless creatures as well? Besides just how harmful is a carpenter bee?

Sure, bad things might happen. A female carpenter bee could sting you. But a person generally has to grab one to be stung; they rarely if ever attack like bumblebees, wasps, or hornets defending their nest. Could burrowing cause structural damage that might weaken a porch roof? Maybe. But since carpenter bees will readily use holes that have already been made instead of creating new ones, demolishing an entire structure would take a lot of burrowing over quite a long time. Our porch may eventually fall down from the annual attack of the carpenter bees, but if it does we will have gotten considerable entertainment from these fascinating creatures for many, many years.

To me, the potential hazards of carpenter bees, and a lot of other animals that evoke paranoia in some people, are not worth controlling with pesticides. That's a high environmental price to pay—not to mention losing the opportunity to watch and hear a live-action show of an industrious pollinator.

What Should You Do If You Find a Baby Bird?

A common question that I receive in spring and early summer is what to do if you find a baby bird on the ground. One writer said she had seen two in her neighborhood recently—one in danger of being run over by a car and the other in danger of being caught by a dog. She wanted to know if there is some group that takes care of homeless birds.

My answer to the first question is a simple one. In most instances the right response is to leave the bird alone. Exceptions arise, but more often than not you do more harm than good by getting involved in a rescue attempt of a little bird.

Exceptions for intervening in a baby bird sighting include removing it from a road where it could get run over (consider your own safety first in regard to traffic). A second reason is to rescue it from an overfed pet dog or cat that is responding to a kill instinct not based on hunger. In either case, simply pick up the bird and release it nearby, and in the second instance put the pet inside. The oft-stated belief that if you touch a baby bird its parents will abandon it has little substance. I know someone who puts recognition bands on dozens of baby bluebirds each year, yet the parents continue to care for them. Abandoning a baby simply because another animal touched it would not be a particularly adaptive response.

The reasons not to rescue a baby bird are far more numerous than the reasons to do so. A little bird could be fluttering around on the ground because it attempted flight too early and launched itself prematurely. It may be being watched from the trees by its parents who are

Q: I recently visited the North Carolina Zoo in Asheboro. Among its many outstanding features is a nice flock of flamingos. While watching them, I asked a zoo worker why some of them were standing on one foot. She said flamingos often do stand on one foot (which was already obvious to me) but she did not know why. I know that flamingos are pink because they eat some kind of shrimp. But does anyone know why flamingos stand on one foot?

A: Flamingos are not the only birds that characteristically stand on one leg. Ducks, geese, and swans often do as well. But flamingos are the poster birds for standing around nonchalantly on one leg for long periods of time, and numerous explanations have been proposed for their doing so. One hypothesis is that letting one leg relax reduces overall muscle fatigue. Another is that having only one leg down at a time minimizes exposure to harmful parasites or fungi that could be in the water. Contact between the bird and noxious organisms would be cut in half if only one leg at a time was in the water. Another reason proposed for a one-legged stance for flamingos is to conserve body heat. But the complete answer probably involves multiple factors in response to a wide array of environmental variables. Scientific findings are often not absolute and some questions about animals do not have a simple answer. Even with proper scientific study we can still be left with uncertainty about something as basic as why a flamingo would choose to stand on one leg.

prepared to teach it to feed. Meanwhile, the parents may be trying to guard against predators that might harm the helpless baby while it is learning to fly. Many people and pets have experienced the wrath of a blue jay or mockingbird protecting its grounded young.

Leaving a baby bird alone after an initial rescue may seem unacceptable to some people because it provides potential for further exposure to a world of predators, cars, and other hazards. But your continued involvement in the process will probably not benefit the species nor, unless you are really good at bird raising, the individual bird. Plus, for most wild US birds a permit is required for keeping them in captivity.

But another reason, an ironic one to be sure, is that you might keep a baby bird alive longer than the parents intended. The bird may be helpless for a reason. Maybe it was intentionally ejected from the nest by the parents or a sibling. Birds have a variety of strategies to deal with limited food resources. One of them is to feed and raise fewer young than the number of eggs they lay. So, in "helping" a baby bird, you may not be fulfilling the preferences of the parents who have assessed that they have too many mouths to feed. They may have intentionally ejected, or allowed a sibling to eject, the baby from the nest. Leave the bird's fate up to nature.

As for groups that care for homeless birds, many wildlife rehabilitation centers specialize in the treatment of large raptors such as eagles, hawks, and owls that have been shot or injured in other ways, some of which are able to be returned to the wild after recuperating. But baby raptors found on the ground should usually be treated like any other baby bird because they may be a product of siblicide (see "Some Birds Take Care of Their Siblings") or the outcome of a parental decision. Most state wildlife agencies can provide contact information for regional rehabilitation centers.

The Mating Game Has Many Rules

One night at the edge of a swamp I watched a big salamander creep along the muddy bottom in search of I know not what. The animal was a greater siren, one of the largest salamanders in the world. The one I watched that night was more than two feet long. I stood on the bank for several minutes, following its movements with my flashlight.

I would have kept watching even if it had just been looking for food, but my hope was that it would encounter another greater siren of the opposite sex. Why? Because despite the immense size of this native salamander of the Southeast, few people see it in the wild, and ecologists know next to nothing about its breeding habits. I was hoping to be the first to see sirens breed.

Understanding the mating systems and social structure of animals is an intriguing area of ecological research. Mating among animals is by no means a random, haphazard process. Within each species, mating strategies and patterns—often highly complex and intricate—have been shaped and molded throughout the species' evolutionary history. Each year our knowledge across the spectrum of species is increased, and the patterns take shape.

Some species, like Canada geese, are relatively monogamous, often remaining with the same mate for life. Male wild donkeys may keep a harem of females and physically prevent younger males from mating with them. The timing of mating also varies among species. Most frogs breed at night. Most birds breed during the day. Zebras breed year round. Garter snakes mate in the spring. Canebrake rattlesnakes look for mates in the fall.

One challenge for ecologists is to collect information about when, where, and how mating occurs, and then find out if and how parental care is expressed. These ecological voyeurs try to determine the social structure of animal populations and unravel the complexities and relationships among individuals. Each new discovery adds insight into our understanding of the relationships between the ecology, evolution, and breeding pattern of the species.

A study of a tropical frog found in Guyana, South America, provided evidence that an individual animal's actions may be best for propagation of that individual's genes but are not necessarily what's best for the species. In this species the female frogs vary in body size and, if given a choice, will select male mating partners that are about four-fifths their own length. Experiments revealed that this size ratio of female to male produced the highest fertilization of a female's eggs. During mating, a male frog clasps a female and releases sperm while she deposits eggs in the water. A male larger than its female mate fertilized significantly fewer eggs because he was not in a proper position for the sperm to reach all of them. Males smaller than the optimal size did not have an adequate sperm supply to fertilize all of the eggs. So to maximize egg fertilization and have the highest reproductive success, a female needs a mate of the proper size, about 80 percent of her length.

However, the male frogs of this species have their own agenda. Larger males can physically displace smaller ones chosen by the female. Thus, a larger one can end up mating with her instead. These bullies of the frog world fertilize a lower percentage of the female's eggs. But they produce more of their own offspring—and pass on more of their own genes—than if they had not mated at all.

In certain species, smaller male frogs sometimes successfully mate by being sneaky. When a male gives a mating call to attract the female, a smaller male may remain quiet but be alert for an approaching female. When a female passes by on her way to check out the calling male, a small, silent satellite male may intercept her and mate. Such competition between males can reduce the reproductive success of some females but ensure the propagation of a particular male's genes.

Every animal has a mating story. However, in spite of our having discovered enough about breeding systems to identify patterns and classify them into categories, the basic observations of individual behaviors and strategies remain to be discovered for most species. Without question, mating patterns exist of which we are still unaware. One day, when someone adds the siren breeding story to the list, we may discover that one of our native wetland species has a fascinating story of its own.

Spring Is Also a Time for Making New Year's Resolutions

We should all make New Year's resolutions in the spring. The vernal equinox, the first day of spring, marks the beginning of a new year. The time to start anew is upon us.

What should you do about it? One suggestion is to make a resolution to go to a library, reputable website, or bookstore and read about the ecology of a group of plants or animals that you know little about. Another is to walk through the neighborhood, park, or woods and see what you can see on your own. Every living thing you encounter will be involved in the process of trying to eat, stay alive, and reproduce.

Spring marks the beginning of the reproductive season for countless wildlife species, both plants and animals. Some breed in other seasons, but in the temperate zones, the majority of species mate in the spring. The strategies employed are myriad: some are obvious; others, more secretive; all, intriguing.

Most species are also making their annual effort to start growing again. New vegetation appears; animals eat the new vegetation and other animals. The stage is full in the spring. If you observe closely at this time of year, you will invariably hear, see, or smell something marvelous.

Each time you hear a bird or insect call or smell a flower from afar, ask yourself why. Is the fragrant flower trying to lure a pollinator? Why are insects attracted to one kind of flower but not another? Is the bird telling others of its kind to come closer? Or is it telling them to stay away? Spring is an ideal time to take a close look at nature and ponder its miracles.

My observations this spring start before dawn with the sound of a mockingbird that directs its musical repertoire toward the bedroom window. The dozens of other bird species that sing in the spring soon join in. Make a resolution to take a dawn walk to hear an astounding symphony. Don't worry about not being able to identify each call. Just enjoy the sounds.

Q: I live near a small lake where I often see black, beetle-like bugs swimming in circles on the surface near shore. What are these bugs and what is their ecology?

A: They are whirligig beetles. Like other insects, they have six legs, but at least one pair has an unusual function: they serve as paddles or oars. More than seven hundred species of whirligigs are found worldwide, so variability in their morphology, ecology, and life cycles would be expected. But several general statements can be made that hold true for most species likely to be observed in freshwater habitats in North America.

Whirligig beetles look like a fast-moving carnival ride when you first approach them from the bank or in a boat, as they go circling and crisscrossing on top of the water. They generally settle down if you remain motionless a short while. The rapid activity is presumably a form of predator avoidance that ceases once they think you are not a threat. Whirligigs mainly use their hind legs to propel themselves through the water. They have large eyes that are divided between top and bottom. The upper half of the eye can see objects out of the water whereas the lower half is adapted to see below the water's surface without distortion. In Alabama, where I grew up, they were sometimes called watermelon bugs First, because they look like a bunch of large watermelon seeds twirling around on top of the water and second because some have a pungent, melon-like smell when picked up (if you can ever catch one).

Resolve to walk around your neighborhood and find ten different plants in flower. To make it more of a challenge, since flowers will be everywhere, try to find the smallest flowering plants you can. Look on the ground. You may be amazed at the number of flowers beneath your feet.

Resolve to find ten different kinds of insects engaged in pollinating the more-obvious flowers. If you are lucky, and can see the smallest insects, you might be able to find ten on a single cluster of flowers.

Make a resolution to watch an insect for fifteen minutes to see where it is going and why. A flying species would probably not be the best choice, but ants, caterpillars, and many beetles walk on the ground or tree trunks. You will find that these animals do not blunder around in the world randomly. Each has a mission. See if you can figure out what it is and learn to appreciate that every single species has its place in the environment.

Watch a spider build its web. At this time of year many spiders prepare for the abundance of insect food that will soon be filling the air. Web making is an intricate art that can hold even a young child's attention. Finding a web in progress might be difficult, but now is the time to look. And if you are looking for spiders building webs you will surely find something of ecological interest, even if you don't find a spider.

Observing plants and insects is a suitable exercise for biology classes. Teachers invariably find that observing living nature is of great interest to most students. If the observations are done in the schoolyard itself, students can benefit from each other's findings. And if your children are not in biology class this year, teach them to take a close look at nature close to home. Teach them to develop an appreciation of nature that will last a lifetime.

If you don't have children at home, take a neighborhood child on a nature walk.

If you failed to make your New Year's resolutions two and half months ago, use the excuse that you were waiting for the real New Year to begin and make them now. Spring will provide many chances to see some fascinating biological performances, and you will increase your own appreciation of the natural world around you.

Ecological Lessons Are All around Us

The prospects look good for something productive finally coming out of my workshop. For the last week, two Carolina wrens have spent each day bringing tasty insects to their babies through an open window I forgot to close last month. If the babies fledge, the parents will have accomplished more in the workshop in a few days than I have in several years. To clarify one point, the shop is really just a small room with shelves where I keep hammer, pliers, and screwdrivers. Hence, it is simply "called" a workshop, so I have no shame that a pair of parenting wrens will outproduce me.

These birds are truly amazing. Such industry. Continuously one or the other arrives at the windowsill with baby food. The first few times each favored me with a look of annoyance suitable for an intruder such as myself. The look from an aggravated wren with a bug in its beak can be disarming. I see why wrens are so popular with people who get to know them.

These wrens make a powerful ecological statement about environmental attitudes. That is, the strongest support for a species or a habitat is derived through familiarity. I have become acquainted with and fond of the two little birds as I've worked outside at a table beside the open window. I saw them building the nest last month, in an old hat lying upside down on a top shelf. Because of my growing attachment to the wren family, should a rat snake decide to visit the workshop for a supper of wrens, I would feel compelled to remove the snake—despite being a herpetologist who appreciates snakes. Lesson: increasing someone's awareness of a species leads to a sense of ownership and protectiveness.

Last week I saw another neighborhood phenomenon that left me staring for several minutes. Early one morning I noticed our dog was entranced with something in the front yard as he watched through a glass door pane. A gray fox was casually strolling around the yard, picking up acorns and eating them. The fox did not once look furtively around to see what might attack it. In fact, its unconcerned attitude, before it finally ambled across the street into a neighbor's yard, was notable. In the spirit of looking for meaning and significance in both unusual and commonplace observations of any plant or animal, I pondered why a fox would forage in such a lackadaisical manner in broad daylight in a front yard. After a few moments' thought I had the answer: we now have leash laws that keep free-ranging dogs off the streets (and out of front yards). A fox no longer has to keep on guard, constantly being ready to escape or fight. Why not stroll around in our front yard and enjoy an acorn or two?

Another Discovery Channel moment came when I glanced toward the top surface of the vine-covered stump of an old oak tree and realized that two enormous lizards called broad-headed skinks were basking side by side. The bigger one was a male, with a shiny brown body and bright red head. The heads of males turn red during the spring courtship period. The female was drabber, but young ones, male and female alike, have metallic blue tails and bright yellow stripes. The female will lay eggs in a nest in a hole in the stump. Lesson: remove trees in the yard as necessary for safety but leave the stumps to serve as homes and habitats for a variety of critters.

Spring is the ideal time to look closely at the sides of trees, in heavy vegetation, or in dark places to enjoy an endless parade of tiny insects and spiders. Check out the plants in your yard and neighborhood in a search for the smallest flowers you can find. Chances are you'll encounter some plants and animals you have never seen before. Be careful what you touch, but getting up close and familiar with living things is the best way to appreciate nature.

Also, don't forget the lesson about leaving workshop windows open and turning an old hat upside down.

St. Patrick's Day Means Snakes
Are on the Move

Two things related to March bring snakes to mind. The first is St. Patrick's Day, March 17. The saint is credited with driving the snakes out of Ireland. The fact that snakes have never lived on that cold island in no way discourages people from making a connection between St. Patrick and snakes.

The spring equinox also occurs in March, around the third week. Days get progressively longer than nights, and both get warmer. Spring officially arrives. Most plants and animals, including snakes, respond to these changes. All US snakes become more active and more evident, hence a word in behalf of this fascinating yet bullied wildlife is timely and always warranted.

Many people today have an awareness and concern about the welfare of natural environments and wildlife, even snakes, and their right to exist in the natural world. Snakes actually serve as a barometer of environmental attitudes of people in an area. An ecologically educated community accepts native snakes as an integral component of natural environments.

Most US snakes are harmless and all are highly overrated as a threat to humans; the chances of receiving a serious bite from a venomous snake are extremely low. The average American probably sees fewer than two snakes a year, and most people never see even one. To be sure, in every natural habitat in the Southeast, Southwest, Midwest, and Northeast snakes are around—underground, beneath logs and rocks, or perfectly camouflaged in leaves, pine straw, vines, or grass—but each year people pass by many more snakes than they see. If you do see one, what are the chances it will be venomous?

Of the more than 130 species of snakes native to the United States and Canada, the majority are found in the Southwest, and of these, only a little over a dozen are venomous rattlesnakes. More than 60 kinds of snakes are found in the East, of which only 7 are venomous. More than 40 other eastern species of snakes are nonvenomous and harmless. Some of them will not even bite when picked up.

Bites of 3 pit vipers (copperhead, pigmy rattler of the South, and massasauga, a small rattlesnake found from Canada and the Midwest to the Southwest) are rarely lethal to humans. Three larger pit vipers found east of the Mississippi River, the cottonmouth and 2 rattlesnakes, can deliver potentially fatal bites. The largest venomous snake in North America, and the one with the most potent venom in the East, is the eastern diamondback rattlesnake, which can reach a length of over seven feet. Next in size, often over five feet, is the timber rattler, also called canebrake rattler.

The cottonmouth is by far the most common venomous US snake that lives around water. The bite of a cottonmouth can be bad, but the snake's aggressiveness is way overrated. Studies have shown that a high proportion of bites from this species occur when people pick the snake up. Yes, pick the snake up! OK. Think about the cause and effect in that scenario, then see if you can figure out a way to reduce your odds of getting bitten by a cottonmouth. Also, most snakes that people see around water are nonvenomous watersnakes whose first response upon seeing a person is to escape.

The copperhead is the venomous snake most likely

to bite someone who is unaware of the snake's presence. But consider these facts: Copperheads bite more people every year than any other US snake, yet human deaths from the bite are exceedingly rare and all have occurred under unusual circumstances. Copperhead venom is less potent than that of most other venomous snakes, and a bite usually causes minimal damage to the victim. A trip to the hospital or doctor's office is still advisable if you do get bitten.

The coral snake is a venomous species that is entirely different from the six pit vipers. A relative of the cobra, an eastern coral snake can indeed kill an adult human if enough venom is injected. But these multicolored snakes are rarely seen and are unlikely to bite a person unless picked up. If anyone has ever received an accidental bite from an unseen coral snake it was truly a rare event. Perhaps the greatest danger is to children who might pick up a brightly colored red, yellow, and black snake.

Snakes are an important part of natural ecosystems in many parts of the country, and their presence and diversity are indicative of a healthy environment. Snakes deserve our respect, and they need not be feared if we follow certain basic safety procedures and recognize that most snakebites are avoidable. The following are suggestions

Q: A friend of mine said an animal I call a glass snake is actually a lizard. Is this true?

A: Yes. They are sometimes called glass lizards or legless lizards because, like a snake, they have no limbs, front or back. Four species are found in North America. Like other lizards, their long tails will break off if grabbed by a predator or a person. Also, glass lizards have eyelids and ear openings whereas snakes do not.

for anyone who lives in an area where snakes occur, which includes most natural habitats in the continental United States. The suggestions are especially important for people who spend a lot of time in outdoor environments. A few venomous snakes also persist in suburban areas.

1. Know the snakes in your locality. If you are concerned about the potential danger of snakebite, find out which venomous snakes might be around and what they look like by checking out educational websites such as those posted by state wildlife agencies or universities or by consulting a state or regional field guide. Phone apps are also available for snake identification.

2. Use common sense. If you see a snake, observe it from a distance of a few feet, but do not try to catch it or disturb it in any way. Many people have been bitten trying to kill a large venomous snake when they could have simply watched it then walked away while it crawled away. If you are in an area that venomous species are known to inhabit, watch where you step and be careful where you put your hands. Rock ledges and fallen logs are prime real estate for snakes.

3. Wear proper attire. When walking through areas known to have venomous snakes, such as swamps and thick vegetation, the safest approach is to wear long pants and high-topped boots or even snake leggings. Leather shoes are too thick for most snake fangs to penetrate.

4. Keep your car keys and cell phone handy. Having access to a vehicle that can transport a snakebite victim to an emergency care facility and a cell phone to call ahead are smart precautions to take. Of course, such forethought would be helpful in any emergency.

The variables associated with whether a bite is serious or minor are numerous. But venomous snakebites are rare. One important fact that needs to be reemphasized is that a high proportion of US bites occur when someone picks up the snake. In those rare instances when the person did not see the snake until too late to avoid being bitten, up to half, perhaps more, of the bites are "dry bites." That means no venom was injected. The potency and amount of venom injected and the tendency of the snake to strike are variables that depend on the species of snakes. For example, an eastern diamondback has venom ten times as potent and can deliver a much higher quantity than a copperhead can. However, a copperhead is far more likely to strike at a person than is a rattlesnake. Dangerous snakebites are uncommon and most can be avoided. With a few simple precautions you can get outside and enjoy the snakes along with the rest of nature.

Children should be taught never to pick up any snake without supervision by a knowledgeable adult. But they, as well as adults, should learn to enjoy snakes by watching them. It's more fascinating and a lot safer than trying to kill them. In any case, snakes will be around one way or another from the spring equinox until the winter solstice, and St. Patrick is no longer around to do anything about it.

Summer

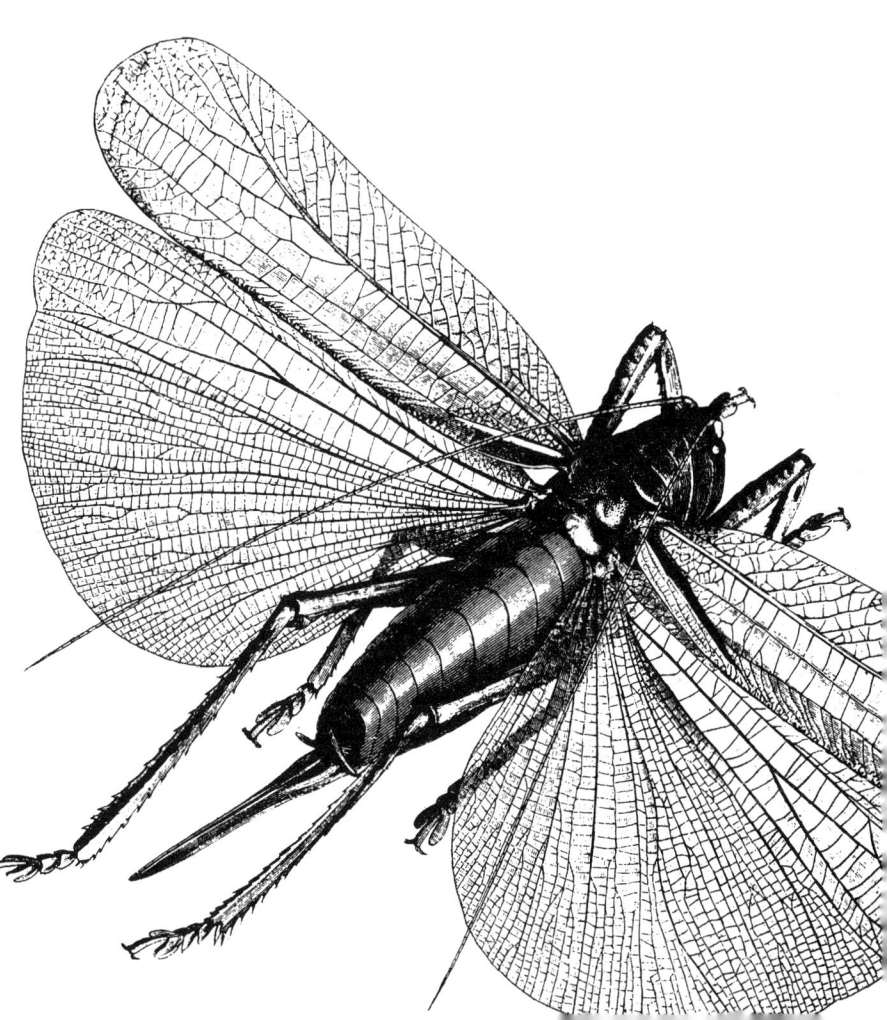

What Can We Learn from Cicada Killers?

We watched from our back porch as a killer ascended to the top of a twenty-foot tree, carrying its defenseless victim. We knew what the outcome would be. The target of the earlier attack would be buried alive in our backyard or our neighbor's, while the perp left the scene unscathed, never to be brought to justice. We could have stopped the flagrant act that was clearly going to end unhappily for the victim, but no one wanted to intervene and interrupt such a fascinating natural phenomenon.

We were watching a female cicada killer, one of the giant black-and-yellow solitary wasps native to the United States, that had stung and paralyzed a large cicada. Cicada killers do not live in nests like some wasps. Instead, each female is a free agent who digs a catacomb-like burrow underground, ready for development of her young from egg to larva to cocoon-building pupa to emerging adult. The process involves a female, which can be more than one and a half inches long, finding a full-grown cicada (which may be almost as big as she is), giving it a paralyzing sting, and then carrying it to the burrow. They really should be called cicada paralyzers.

The cicada ultimately gets carried to the prepared tunnel and deposited in a side pocket the cicada killer has dug, a sort of cave. The mother then deposits an egg on the permanently helpless cicada. When the egg hatches, the larva enters the still-living cicada's body, and then proceeds to get the nourishment it needs to complete the development process. Skip the sugar-coated explanation:

the larva literally eats the cicada alive over a several day period. After fattening up, the larva then turns into a plump pupa that spins a cocoon and stays underground in its own little crypt for a year or so. It emerges in the summer ready to go on a cicada quest if it's a female or to look for female cicada killers if it's a male. As is true of many animal species, the males emerge from the ground and become active earlier in the summer than females.

But before all this teenage cicada killer development can happen, the soon-to-be mother faces the dilemma of locating a cicada. Step one is finding her prey during the daytime. When she does, she is an awesome airborne attacker, being able to catch a cicada and bring it into submission with a sting in midflight. Considering what it's like and how painful it is when a honeybee or wasp stings creatures like ourselves that are a thousand times larger, what is it like to be a defenseless cicada? Would it be like a person being jabbed with an eight-foot long, two-inch diameter needle with an oil drum full of venom? Instant paralysis is a plausible result.

The cicada killer's next problem is getting the cicada to the burrow, which after the female's search may be a football-field length away. Dragging it over the ground is one approach, with the female using her back legs to hold the immobile cicada under her own body while using her other four legs to walk. But when the burrow is a long distance away, these enterprising insects have another strategy. Climb a tree with the cicada quarry in tow and then launch and glide in the direction of the waiting cicada killer nursery. This is what the one we watched was doing as it walked rapidly up the tree trunk to get a higher vantage point and glide toward the burrow.

Meanwhile, what do the males do while all of this cicada killing is going on? They, too, have interesting behaviors to observe and probably ones you are more likely to see. Male cicada killers will even attack humans, but we really do not have to worry because the males do not have stingers. Like many animals, including humans, these enormous wasps have powerful territorial instincts.

These warlike black-and-yellow insects look like they might deliver a serious sting but are generally harmless to humans—the females because they only sting people if they are picked up or stepped on; the males because they do not have stingers.

Female cicada killers spend most of their time looking for cicadas. The males spend their efforts preparing to meet and mate with females. Adult males emerge from the ground in summer before the females and set up mating territories, which they defend against other males. The territories are established in areas where female cicada killers will be emerging from the ground. Upon emergence, the females normally mate with a territorial male.

Male cicada killers confront other males that enter their territory. Using intimidation as their primary weapon, males will sometimes even defend their territory from other animals, including people who enter the area. Despite their imposing appearance, if a male cicada killer chases you, simply stop and enjoy its antics. Or if you prefer such sport, run, and it will chase you.

A clever study conducted by Perri K. Eason and Gary A. Cobbs at Northeast Louisiana University (now University of Louisiana at Monroe) and Kristin G. Trinca of the University of Louisville has shown that when boundaries are clearly marked, the males have fewer problems with trespassers. The research findings also illustrate once again that any animal species, even a lowly insect, can have fascinating behavior. Researchers conducted field experiments with cicada killers to confirm previous anecdotal reports that naturally occurring landmarks are used by some species to define territorial boundaries. To test the importance of visual landmarks in territorial behavior, investigators caught individual males and painted unique patterns of colored dots on their bodies. After releasing the insects in a flat, grassy area with no obvious landmarks, they determined by field observations the size and shape of each defended territory, generally occupied by a single male that was identifiable by its paint pattern.

The researchers then placed thirty three-foot-long wooden dowels horizontally on the lawn in a random pattern to serve as tangible landmarks in the otherwise homogeneous habitat. Upon remapping the territories of the wasps, the investigators found that by the next day forty-two out of sixty-two territories had been redefined within the study area, with the dowels being used as boundary markers. No male's territory crossed a dowel into the territory of another male, indicating that these landmarks were observed as the boundaries by males on either side.

Further observation revealed an interesting phenomenon among wasps that were defending boundaries where a dowel could be used as a landmark on two sides and where no clear landmark was available on the other two sides. These males spent more than six times as long during the day defending the sides with no landmarks as they spent defending the sides where dowels were present.

In the experiment, dowels represented natural landmarks. One conclusion offered by the investigators was that the use of obvious natural landmarks to define territorial boundaries could have evolved because of the reduction in costs of territorial defense. That is, establishing a territory in a homogeneous habitat is not energy efficient for a cicada killer because of the constant patrolling and vigilance necessary. When boundaries are distinct, the wasp can be more resourceful.

The studies with male cicada killers suggest that some insects apparently would agree with the neighbor in Robert Frost's poem: "Good fences make good neighbors." The reasoning behind this homespun philosophy is that the powerful territorial instincts of many animals are more efficiently managed when boundaries are distinct. Whatever Frost thought about the stone boundary in "Mending Wall," we think he'd agree that for these insects good fences result in neighbors who know where the boundaries are.

Color Means a Lot in Ecology

What color is the Fourth of July? Most older Americans probably think of red, white, and blue. Some children may think of bright yellow and other exploding parts of the color spectrum associated with fireworks. Halloween is orange and black; Thanksgiving is green and brown (in my mind) for reasons I am unsure of other than those are the colors displayed on bulletin boards in autumn. Christmas is red and green. And Valentine's Day is red and white. Color is clearly an important component of our lives. It is also a defining part of the lives of many plants and animals and frequently tells an environmental story.

The array of color patterns in plants and animals can be an inducement for anyone to become an amateur ecologist. All you have to do is be interested in observing nature and the outdoor world of plants and animals. In the simplest terms, all you have to do is observe in a questioning manner. When you see a plant or animal, ask yourself why it is one color instead of another.

The fact that plants with chlorophyll are green or that dead limbs and vegetation are brown may not be particularly intriguing to most people. But more striking colors, of obvious significance to the existence of the plant or animal possessing them, set the stage for environmental mystery. And not-so-colorful animals can prompt the curious to wonder why an animal sports a particular pattern of stripes or spots. Even ecologists themselves do not always agree on the answers.

The reasons for some color patterns seem apparent.

For example, some species of male birds develop bright plumage in the winter and spring, whereas the females remain drab. This is the period of courtship and, as most birds can distinguish color, the males use their plumage to attract and impress females. Male American gold-finches turn bright yellow during the mating season. Redwing blackbird males display bright red and yellow epaulets on their wings. The females of redwing black-birds are an unimaginative brown, and the female gold-finch is a much paler yellow than the male. The breed-ing colors of the males in some species are also used to threaten or intimidate other males. The ready distinction between males and females assures that courtship efforts or acts of aggression are not expended on the wrong sex.

The mammals represent a group for which a general question can be asked about color: why are most mam-mals some shade of black, brown, or white? Birds, but-terflies, and flowers come in an endless array of yellows, reds, and blues. But except for a few of the primates, mammals seem to be stuck with varying shades of dull.

Ecologists generally consider that the various shades of brown seen in many mammals are for camouflage. Camouflage is important for both predator and prey spe-cies because, in either case, the animal does not want to be seen. Thus white-tailed deer and mountain lions are brown, a color that blends in with forest habitats. The spotted coats of adult leopards and baby deer help them hide from the eyes of other animals, prey in the case of leopards, predators in the case of fawns. Although their reasons for having spots are different, both animals have clearly adapted to blend into particular habitats.

Numerous examples exist for which the environ-ment dictates the color pattern, whether for protec-tion of a prey species or for stealth in a predator. Some

Q: Why do the anole lizards we see that can change from brown to green sometimes have a bright red throat?

A: Males of many species of anoles use a throat fan, or dewlap, to challenge other males, and sometimes even people. The dewlap is typically bright red in the native green anole and yellow, orange, or a combination of colors in some of the introduced anoles now found in southern Florida. The display of the dewlap is often accompanied by the male lizard doing push-ups and bobbing its head. Next time you see one with a noticeably expanded throat, hang around and see if he will put on a show.

species, such as the snowshoe hare and Arctic fox, actually change their coat from brown in summer to white in winter in keeping with seasonal changes in color. The brightly colored rump region of male baboons is one of the most obvious displays of color among the mammals, except for hair color among some of today's teenagers. One suggestion for why mammals have few displays of bright color is that color perception among most of them is not the same as for birds and primates. If this is true, the common function of color to attract or dispel others of the same species serves no role in most mammals.

Many fish are also distinctive in their color differences between the sexes. The brilliantly colored darters of the southeastern streams are dramatic in their contrast between males and females. In the male Christmas darter of Georgia and South Carolina, bright red and green bars are present during the breeding season. Males of the redband darter of Tennessee sport bright blue and red orange on their fins. Female darters are generally drabber in appearance.

One problem for ecologists comes in trying to explain black-and-white color patterns in mammals. Most species with black-and-white fur have special traits. The prime example in North America is the skunk. Presumably the black-and-white contrast is not for protective camouflage but as a warning signal to other species. Even a bobcat without color perception as we know it can see the contrasting black and white and learn not to tangle with what might otherwise look like an easy meal.

Some black-and-white color patterns of mammals are less easy to explain. And ecologists often do not agree on what the correct explanation is. For example, at least fourteen different hypotheses have been proposed for why zebras have black-and-white stripes. One suggestion is that the striped pattern creates moving vertical lines difficult for a predator, such as a lion, to focus on. Thus the herd can sometimes escape before an individual has been singled out for capture by a confused cat. Another is that zebras once lived in partially shaded forested areas where the stripes served as camouflage. The stripes in this case are considered holdovers from an earlier time. Whether either hypothesis, or a dozen others, has any validity is uncertain. Questioning why the pattern exists is what is important. You may even be able to offer an explanation for zebras' coloration that has not been thought of.

Color is also used as a lure in some animal species. Baby copperheads, cottonmouths, and pigmy rattlesnakes have bright yellow tails that are waved enticingly in the presence of small frogs or lizards. Because the rest of the snake is well camouflaged in dead leaves, the frog or lizard becomes prey when focusing its attention on the tail and mistaking it for something to eat.

Flash colors are a special use of color for defense

among some animals. For example, the gray treefrog is a perfectly camouflaged creature when sitting on an oak tree or other drab background. When a gray treefrog is pursued by a bird that intends to make a meal of it, the frog jumps and displays bright yellow underparts. Upon landing on a tree and tucking in its legs, the frog blends into the background. The bird, meanwhile, is in search of something yellow that cannot be found.

Plants also use color to great advantage, various forms of color advertising being the most apparent. Brightly colored flowers attract insects that are essential for pollination in some species. And few plants can be accused of false advertising as the insect lured to a flower is usually treated to nectar. Bright red or yellow berries that attract birds such as cedar waxwings offer a meal for the bird and assure that the enclosed seed will later be deposited in another area. One point to consider is that insects that pollinate what we perceive as white flowers may actually be seeing a different, possibly more distinctive, color in a different part of the color spectrum, such as ultraviolet, that we are not capable of seeing.

One color phenomenon, albinism, is not a product of the natural environment of plants and animals. Albinism is the expression of a genetic condition that can be inherited, although neither parent need be an albino itself. An albino is incapable of producing the pigments that normally give color to hair, skin, feathers, and other surface tissues. Because of this abnormal condition, survival in the wild is a difficult struggle. Ironically, laboratory white rats and mice are the most successful and plentiful albinos in the world and serve to demonstrate how human intervention in the natural world can have an enormous impact on what constitutes success in a particular environment.

The world around us is full of color displays by plants and animals. Horticultural plants and domestic animals are less interesting ecologically because many have been bred to produce certain color patterns. But the colors of native species give a special message about the lifestyle of the organism. Somewhere in the past or present is an explanation for the colors we see, even if that explanation is not obvious to us. Next time you see a moth with pink-and-yellow wings, a brown chipmunk with white stripes, or a trumpet vine with orange flowers, ask yourself, why is it this color instead of another?

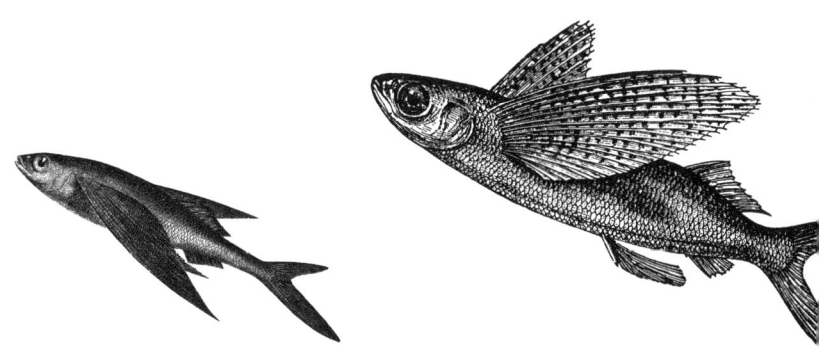

Questions about Alligators Never End:
Part 1

How big do alligators get, how long do they live, and how fast can they run when chasing a person? These questions are asked every year, especially by residents living in regions where alligators occur naturally. Alligators, the largest of our native resident reptiles, inspire so many questions that the most straightforward way to write about them is to simply serve up the questions along with the answers.

Q. My nephew asked me if alligators can jump over a fence. I thought I heard once that they could. Also, do alligators have the ability to jump out of the water the way fish do, and if so, how high?

A. Alligators cannot jump over fences from the ground, but I have seen alligators up to six feet long climb over both chicken wire and chain-link fences. I'm not sure how high they can climb to get over a fence but certainly several feet. I saw crocodiles on the Adelaide River in Australia jump to grab meat from the end of a stick ten feet above the water. I have never seen alligators do this, but since they have a body shape similar to a crocodile and have a flat tail, I assume that they might be able to, at least for a few feet.

Q. Can you please help us with some information? We need to know what baby alligators and baby crocodiles are called.

A. They are referred to as hatchlings when first born and juveniles later, but they have no other special name. I have asked a few herpetologists to suggest a name but haven't gotten an answer from anyone yet. I suppose "gatorling" might work for alligators and "crockling" for crocodiles. But until those terms catch on, baby alligators and baby crocodiles will have to do.

Q. Our daughter said she heard that a male alligator's eyes are red, whereas a female's are not. Is this true?

A. The eyes of male and female alligators are yellow with a black pupil. What your daughter may be referring to is eyeshine, the reflection from some animals' eyes in the dark. In alligators, the reflected color can range from deep ruby red to orange to yellow. Under some conditions the reflection is visible more than a hundred yards away. The eyes of big alligators seem to me to appear deep red whereas smaller ones are more often yellow, but this can vary. Male alligators get bigger than females, but I am not aware of any inherent difference between the eye color of males and females.

Q. A friend said she heard male alligators bellowing somewhere in southern Alabama during a trip in early summer. Is this true? Could the noise she heard simply have been bullfrogs?

A. It could have been either, depending on when and where she was. Alligators do make deep, resonant vocalizations, especially in the spring when they mate and could still be doing so in the summer. An adult alligator makes a rumbling sound, much louder and more guttural than a bullfrog, that you can practically feel if you are standing nearby. I have heard bellowing from a large male alligator (about thirteen feet) and a female (nine feet) that live in a pond in South Carolina; both the male and female may bellow for several days during spring. My impression is that the female is responding to the male's bellows and that she is actually louder. But either would make a bullfrog's call pale in comparison.

Q. I belong to a water-ski association near Charleston, South Carolina. We occasionally see gators in the area, and after an alligator attack was reported in Florida some of the skiers are a little anxious about getting in the water. No one in our group feeds the gators, and our members are skiing enthusiasts who enjoy water-skiing more than they fear gators, but any information you can give us to make our sport safer would be appreciated.

A. News stories about alligator attacks always generate many questions, and your concerns are understandable, but the threat is minimal. I have worked around alligators for many years, and the only bites or bruises my colleagues and I have experienced have come while actually capturing a gator. Although the bizarre is always possible with wild animals, I would not hesitate to water-ski merely because alligators are present. The dangers on the highway leading to your ski area are a thousand times greater than the dangers from the reptiles in the water. But traffic accidents are commonplace, whereas experiences with potentially harmful wild animals are in the realm of the unknown for most people.

One important point: large alligators should be wary of the presence of humans where you are skiing and stay some distance away. If they are not and you see an alligator swimming toward you while you are on the shore, it is likely expecting a handout of food because people have imprudently fed the gators at your ski area in the past. Be extra alert if one approaches the boat or swims toward a person in the water. You mentioned that water-skiers in your group do not feed the gators, but other people might. Alligators that have been illegally fed from the shore or boats are attracted to people and are expecting to get something to eat, which regrettably could end up being a person. Although it is of little comfort to someone who has been attacked by an alligator, such incidents are extremely rare.

Q. Are large alligators more aggressive than smaller ones? What is the most dangerous size alligator to someone who is on land? It seems like smaller ones could run faster.

A. As a general rule, the most dangerous size alligator to find in the wild is a newborn baby. Many adult females (which are smaller than the males but can reach lengths of more than nine feet) will vigorously defend their nests and babies. A baby alligator may let someone walking along shore pick it up without trying to escape, but when it is scared it will start making a squeaking sound, which brings the mother gator to protect it. A mother alligator will come up on land with mouth open to chase a person away. Her objective is to frighten away the intruder, and she typically will not continue in pursuit once a person retreats. The behavior is perceived as highly aggressive, but in reality it is strictly a defensive measure to protect her young. Aside from parental protection by females, attacks on humans generally involve situations in which alligators have been fed by people and thus have developed atypical behavior.

Q. Do baby alligators really call their mothers the way birds do?

A. Most biologists consider birds to be the closest relatives of alligators and crocodiles evolutionarily, and many similarities between the groups are apparent. Not only do baby alligators call their mothers, they start doing so before they even leave the nest. Mother alligators lay their eggs on land inside large mounds built of dirt and vegetation. At the time of hatching, the babies begin making yelping sounds that attract the mother to the nest. She digs into the nest, opens eggs with her teeth if necessary to let the babies free, and will even carry the young to the water in her mouth.

Once in the water, baby alligators stay in the vicinity of their mother, and when one feels threatened, it will make gulping or yelping sounds that can be heard several feet away. A mother alligator will defend her nest and young from predators and will investigate when she hears a baby in distress. She will attack another animal, including a person, that appears to be a threat to her babies.

Male alligators are aggressive toward other males during the mating season and perhaps other times, but I have never seen one be aggressive toward a person. However, it is conceivable that a territorial male alligator could temporarily mistake someone swimming as a competitor male, or perhaps as some other invader of its territory, and attack. For most people, common sense would suggest that any animal bigger than we are with a big mouth and lots of teeth should be viewed with a cautious eye, and if you know one is in the water where you want to swim, my suggestion is to give it the right-of-way.

Q. Are there any defensive moves to make if attacked by an alligator?

A. On land, back away, then run if an alligator comes out of the water toward you. If you are in the water, try to get to shore. As a last resort, fight back by hitting the animal in the face, even though the effort may seem futile. Yelling for help would be a natural response that should be employed immediately.

Q. When is alligator mating season?

A. The mating season in most parts of the American alligator's native range usually starts in early spring with male territorial behavior toward other males and courtship of females. The females lay eggs in early summer in a large nest they build on shore from mud and vegetation. The babies hatch out in late summer or early fall. When the babies begin to hatch they sometimes make little grunting sounds and the mother will come out of the water and dig open the nest. I have watched a mother alligator pick up eggs in her mouth, crack them open so the babies can get out, and then carry as many as three babies at a time down to the water until all were there. An awesome spectacle. The female is protective of her young for a year or more.

Q. Are alligators a type of lizard? Is it true that alligators are kin to birds?

A. Alligators are reptiles that are most closely related to crocodiles. Other major groups of reptiles are the turtles, snakes, and lizards. Technically, alligators are no

Q: While walking in the woods in north Florida we found a box turtle that had been partially eaten by a predator. Inside her body were four oblong white eggs with firm shells but also lots of smaller, round yellow eggs. Is spring when box turtles lay their eggs? Is it possible to hatch the eggs? Would I need an incubator of some sort?

A: Box turtles lay their eggs in the spring or summer, as do all but two of the more than fifty species of US turtles. And, yes, you might be able to hatch the four shelled eggs. The shelling process within a female turtle generally takes from a few hours up to a day or more. The calcium shell layer of an egg forms around one of the bright yellow follicles (yolk) and becomes leathery. The female may lay the clutch right away or hold them for several days while waiting for suitable environmental conditions of temperature and rainfall. All US turtles lay their eggs on land by digging a hole in the ground, depositing the eggs, and then covering the nest with soil. You can keep turtle eggs on moist paper towels to hatch them; do not let them dry out or get so wet they get moldy.

more closely related to lizards than they are to snakes. Although alligators are shaped somewhat like lizards, they are only distantly related to them. Scientists consider birds to be the closest living relatives of alligators and crocodiles.

Q. Where do alligators live in the United States? I have heard that they can be found in Oklahoma and Arkansas.

A. Alligators occur naturally on the Atlantic and Gulf Coastal Plain from North Carolina south around Florida and into Texas. Natural populations of alligators can also be found in about half of Arkansas and extreme southeast Oklahoma. They occur throughout Florida and Louisiana. Incidentally, another species of alligator is native to China; the Chinese alligator is bordering on extinction in the wild.

Q. How fast can an alligator move on land? Is it possible for a person to outrun one?

A. Alligators can run about as fast as a person for a few feet but cannot sustain that speed for very long. I have never encountered an irate alligator (always a female guarding a nest or babies) that I could not outrun, which is why I am able to answer your question.

Questions about Alligators Never End: Part 2

Q. How big was the largest alligator ever captured?

A. In a study done in Florida from 1977 to 1993, the largest male alligator was fourteen feet long and the largest female, ten feet, two inches. An alligator that was killed and left in a Louisiana marsh in the early 1900s was estimated to be over nineteen feet in length, but that record has long been disputed. The largest size verified for an alligator based on a statistical analysis of skulls, skins, and live animals was slightly less than fifteen feet, and a large male killed during alligator hunting season in 2015 in Alabama was more than fifteen feet long.

Is the absence of the giants reported from yesteryear a consequence of the elimination of most older, larger individuals during the early part of the century? Or were the extreme sizes reported in the past a result of unintentional misreporting or mismeasurement? Whatever the true maximum size ever attained by an alligator may be, twelve- to fourteen-foot individuals weighing more than a quarter of a ton are around today.

Q. We live in Florida alongside a canal and lake. Alligators in these waters are not at all intimidated by the presence of humans. They seem to be attracted to children, often coming within several feet of youngsters playing in the yard. I feel that they are a threat to our children and should be removed from the area. They come right up to our deck and sit in the water, watching us closely. If I walk down our boardwalk along the water, they follow. We have heard accounts of alligators coming into backyards; several dogs have been taken by an eight-foot-plus gator. Can these gators be removed?

A. The state wildlife department should be contacted for concerns about so-called nuisance alligators. Though your question could have come from someone in any state with alligators, since you are in Florida, contact the Florida Fish and Wildlife Conservation Commission. Meanwhile, some common sense cautions can help prevent problems with the alligators. Do not let children or dogs play near the water. Do not pick up baby alligators (which will begin appearing in late summer or fall). Of critical importance is the admonition not to feed the alligators. You should also caution others not to do so. Someone in your vicinity has probably fed the alligators, which would explain their interest in humans. In addition to creating a nuisance, feeding them is illegal. As with so many animals throughout the world, even those that may be perceived as a nuisance or even a threat to humans, we should keep in mind that we have invaded their home, not vice versa.

Q. Why have there been so many attacks by alligators in Florida?

A. Three women were once eaten by alligators over a several day period in Florida, and many other gator attacks have been confirmed. The answer to why is a simple one. The population density in Florida continues to rise, placing more and more people in association with the largest native reptile in North America. Although each individual case is a personal tragedy, many of the people attacked, bitten, and sometimes killed by alligators placed themselves in harm's way. Alligators are not doing much different from what alligators have always done. People are.

Often people are simply naive: swimming in lakes and canals at night where large alligators are known to live; being new to the region and ignorant about the wildlife that inhabits the region (and always has); walking a dog along a waterway with big alligators (alligators will come out of the water to eat a dog).

Again, not to minimize the personal suffering of families affected, alligators kill fewer humans in a decade than cars kill in a day and dogs or horses do in a year. Why do alligators harm even a few humans? Most injuries are caused when humans have fed the alligator (an illegal activity), invaded its territory (including moving into it permanently), or threatened its young (baby alligators are cute and will allow themselves to be picked up, a bad idea). When humans choose to live in a place where they might come in contact with alligators or any other large animal, wild or domestic, each person should be responsible for knowing where not to tread.

And it goes without saying that it is the responsibility of parents and guardians to watch out for children under their care. You can't expect an alligator to do so.

Q. What do you think about the outcry in Lakeland, Florida, about alligators eating the mute swans?

A. Lakeland, sometimes called the City of Swans, has a flock of two hundred mute swans swimming around in a downtown lake. Lakeland may have to change its name to the City of Gators if a recent trend continues—alligators have eaten almost a dozen of the swans in the past month. Some of the residents have called for removal of the gators. What do I think? I think that mute swans are native to Europe and that alligators, which eat waterfowl, are native to Florida. Alligators, as I said before, are not doing much different from what they ever have. Mute swans are.

Q. My neighbor swears he saw a large alligator swimming in the ocean somewhere along the Charleston, South Carolina, coast at least two hundred yards from shore. Do alligators ever leave freshwater lakes and rivers?

A. Alligators leave freshwater habitats to travel overland during droughts, in search of mates, and to avoid confrontations with larger male alligators. They will enter saltwater habitats on occasion and have even been found a mile or more out to sea. They do not live in the ocean but can tolerate saltwater for hours or maybe even days without a problem.

Q. Is there a sonic sound wave that can scare alligators out of the water?

A. As far as I know, this is not an effective way to make alligators leave the water, but they are able to detect water vibrations through sense organs in the jaws and perhaps would respond by leaving the water. However, if sonic vibrations in the water bothered a gator, it might just lift its jaw out of the water. Meanwhile, other animals in the lake might be affected in a negative way as well.

Q. I've written to numerous agencies and never even received a reply at all about this: what is the justification

Q: What would be the best way to combat an invasion of pythons to other regions, and will the problem be easy to control?

A: Total elimination of pythons by human intervention will probably not be possible in some situations except on a case by case basis. An extended cold winter could eliminate local populations in many areas. Complete control is probably too late because pythons have already become established in southern Florida. A young python could easily hide on construction materials, horticultural items, or other goods and then be transported anywhere in the country. Convincing people who have pet pythons not to release them into the wild would be a big step in preventing further invasion, since that's probably how a lot of them initially got into Florida and other states.

for preserving dangerous reptiles such as the alligator? It's a lame excuse to say vaguely, as a politician might, that it would upset an "ecosystem." Are there specific and documented reasons that the American alligator, in particular, is so protected? Of what real value is it?

A. The answer to this question is incredibly complex, will be received with varying levels of acceptance by different audiences, and involve facets of sociology, psychology, and economics as well as ecology. Some people will never accept any reason as justification for preserving animals such as rattlesnakes, tigers, and sharks that could kill us; other people will always side with the animals.

One question I would ask is "who is defining the reptile as dangerous?" I have dealt with venomous snakes, crocodiles, alligators, and many other such creatures most of my life and would not define them as any more "dangerous" than lawnmowers or electrical outlets. I know dozens of people who have been killed or injured by cars and have read about thousands more—are cars dangerous? Should we get rid of them? Compared to hundreds of other potential hazards we face in everyday life, "nature, red in tooth and claw" poses a minor threat. Few people get injured by wild animals or die from encounters with them.

In essence, I do not believe that any wild animal should be indicted just because we are capable of putting ourselves into a dangerous situation with regard to it or because it has no practical value to us. As I said, I have dealt with potentially dangerous animals for years, and I know that when I have gotten hurt, it was my fault, not theirs. Their value to me has been the excitement and pleasure of knowing they exist.

Another reason for protecting all wildlife species

is that once we declare that a particular species is not worth saving because some people find no value in it, another animal will be next on the list. Do we get rid of whichever species seems to be the biggest nuisance, or the most dangerous, or the least useful at any given moment? Should we get rid of blue jays because they are raucous or channel catfish because they have lateral and dorsal spines that can injure us or squirrels because they raid our bird feeders? And who will decide which species should be the next to go? Finally, some of us, possibly the majority of people living in North America, just plain like wild creatures. That may be the best reason of all not to eliminate them.

A seldom-mentioned reason for maintaining a high diversity of wildlife in the world is that many species keep the human spirit alive by spurring curiosity. Alligators clearly spark people's interest, and the high number and variety of questions people ask about them are positive signs. When people express wonder about any kind of native wildlife, the species in question is stimulating awareness, which helps keep people interested in their surroundings. And the more interested people are in the environment, the more interested they are in protecting it.

Let's Go Out in the Swamp Tonight

The request by my seven-year-old grandson Nick to "go out in the swamp at night" was one any self-respecting granddad would want to honor. We were spending the weekend out at our cabin in the woods and that evening we set out on our excursion.

Sunset was at 8:27 p.m., or so my grandson informed me, having gathered the background information we needed before leaving his house. By 7:00 he had gathered up flashlights and donned a pair of children's chest waders. Nothing like getting ready ahead of time. I took charge at that point and declared that as it was still daylight I would have a cup of coffee before we began our great adventure. When we started our trek an hour later, I had put on rubber hip boots. A lingering twilight filled the sky, which gradually darkened as we walked through two-foot-high ferns over terrain that varied from lush carpets of emerald green sphagnum moss to clear pools of foot-deep water. Flashlights were not really necessary at first, but before long total darkness was upon us. On came our lights.

I led the way, with Nick following a few feet behind. When we came to water, he held onto my back pocket lest he trip or we hit a deeper pool. We stopped occasionally to turn off the flashlights and listen to the sounds of the swamp. "Cool," said Nick when we heard the commanding call of two barred owls challenging each other for dominion over that part of the swamp. We heard a cricket frog, whose call Nick likened to someone hitting two marbles together. Then we heard a loud chorus of green treefrogs, their quacking sounds more like ducks than frogs. I suggested we go in search of them.

Q: Where are the most species of frogs and toads found in the United States?

A: Indisputably, the greatest biodiversity of US frogs and toads is in the southeastern states from the Carolinas to Florida, Georgia, and Alabama. Each of those states has at least thirty species of native frogs, which is more than any of the western states.

We walked and waded along, shining our flashlights carefully ahead as I parted vegetation. I was hoping to find a cottonmouth, but with plenty of time to see it and watch it, not step on it. We saw a banded watersnake slither into a dark pool and disappear. We finally reached an area of two-foot-deep standing water where we were surrounded by noisy green treefrogs. None could be seen, but we could hear them all around us. Then Nick saw one calling from the stem of an aquatic plant. He moved stealthily and with the skill of a professional herpetologist grabbed it. We looked it over, agreed that it was a beautiful animal with its brilliant green back and white racing stripes down both sides, and placed it back on the plant. It was now late and time to head back to the cabin.

As we walked on solid ground along a ridge with swamp on both sides, I proposed that we stop one more time and turn off our flashlights. This time we heard the upbeat call of a chuck-will's-widow, a close kin of the whippoorwill. We paused for a minute more, staring at total blackness all around, and then it happened. One of the most amazing insect displays imaginable. Lightning bugs. Hundreds upon hundreds of lightning bugs flashing their signals. But these were not the random

everyday (or night) glimmers of backyard lightning bugs we have all seen. These fireflies were flashing in unison, in total synchrony. One moment the swamp was alight with twinkling bioluminescence. The next instant the inky black night enveloped us. Then the fireflies lit up the swamp again.

Nick whispered a single word: "Awesome." I silently agreed.

Only one species of lightning bug (*Photinus carolinus*) in North America has this synchronous flashing, and the precise biological explanation for why they do it remains a mystery. I have seen the phenomenon three times, each time while in a swamp. You don't have to be a seven-year-old or a scientist to know that this was indeed an awesome sight.

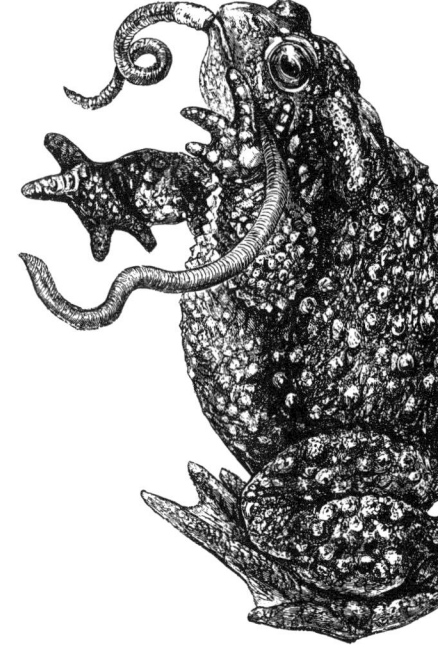

"But Poison Ivy, Lord'll Make You Itch!!"

Before we dug up a gopher tortoise burrow as part of a study, I was one of those people who claim they "never get poison ivy." Three weeks later the itching had almost stopped. The connection between digging in the sand and my ailment is only conjecture, but one hypothesis is that oils were released from the roots that got chopped up. The oils were deposited on the sand where we were crawling around. I can say with certainty that the only places I became infected were where my arms touched the sand. You do not have to touch the leaves to get the blisters on your skin.

Poison ivy and poison oak are the classic outdoor pests of the plant variety. Most botanical field guides distinguish between poison ivy and poison oak. But the differences are subtle and do not really matter in identifying the plants. The two most definitive traits are the presence of three leaves on each stem and a red coloration at the apex where the three leaves connect. Both plants produce the oils that make you itch. Some biologists claim that poison oak is more virulent than poison ivy, but others say this has not been firmly documented. I suspect someone who is itching like crazy from exposure to either plant doesn't care that the other one might be more virulent.

Poison ivy can climb trees as a vine (usually attaching close to the trunk), look like a shrub, or be a single, simple plant. All parts of a poison ivy plant—leaves, stem, fruits, and roots—produce oils that can cause skin irritation in some people. Any bodily contact with the oils can cause a problem, whether from patting a dog that has just walked through poison ivy or touching clothes that have come in contact with the plant. You can even get a case of poison ivy internally by inhaling oil droplets that become airborne in smoke when the plants are burned. You can also get poison ivy in the winter simply by touching the stem or vine, even though the leaves are gone. I know of someone's boyhood experience gone bad after climbing around on an oak tree with a big vine that turned out to be poison ivy.

Nonetheless, after spending thousands of hours at a research ecology lab with people who spend time year-round in swamps, woods, and streams where poison ivy is as common as a household word, few of us ever got a serious case, aside from the aforementioned encounter with the roots in the tortoise burrow. Perhaps ecologists, hunters, and wildlife managers who are in the woods a lot avoid the plant without being aware they are doing so. Most people have no reaction when they casually brush against poison ivy, and as many as a third or more are not sensitive at all. Some people spend their lives around poison ivy without ever having a reaction.

The facts, myths, and disagreements about the properties of poison ivy are legion. You do not spread poison ivy by scratching where it itches, despite what some people say. New blisters and irritated areas can appear more than a week after exposure to the oils, but these merely represent the normal lag time that can occur after initial

Q: What does an ecologist do?

A: Ecologists study the relationships between plants, animals, and their environments. They also teach others to understand and appreciate the natural world we live in and depend on.

contact. You cannot give poison ivy to others, except by bringing them into contact with the oils that are on your own body or your clothes after encountering the plant.

Many forms of wildlife can eat poison ivy without being adversely affected. Dozens of kinds of birds including bobwhite quail eat the fruits, which are clusters of smooth, white berries that appear in late summer. I know of a serious case of poison ivy being contracted by a student who sorted through the stomach contents of a recently killed deer. Among the data he recorded for his research on the diet of deer was that they sometimes eat a lot of poison ivy leaves.

According to my favorite dermatologist, the symptoms of poison ivy are a consequence of the oils causing a contact dermatitis (inflammation of the skin). But she says that superficially similar dermatitis can be caused by numerous other plants and even commercial products we might come in contact with. Among the plants known to cause dermatitis are black walnut trees, red cedar, and fresh okra. Fortunately, most people are not affected by those plants.

The skin irritation—blisters, burning, itching—resulting from poison ivy normally occurs twenty-four to forty-eight hours after contact with any part of a poison ivy plant, and expression of the ailment follows a bell-shaped curve. The most severe symptoms occur midway between a two- and twenty-four-day period. One treatment for relief of the symptoms of a severe case is a steroid such as prednisone. The steroid masks the symptoms even though the body's response to the irritation continues. My dermatologist cautions that some doctors treat with steroids for too short a period. Thus if your reaction is following a twenty-four-day cycle and you take steroids for only seven days, the symptoms could reappear before the peak of irritation has been reached.

The oils produced by poison ivy that make us itch are not directed toward protecting the plant from humans. The fact that some people experience dermatitis from an encounter with the plant is purely incidental. However, some biological products of certain plants and animals are specifically designed to protect them from harm by other species. The arsenals used by plants and animals throughout the world in waging chemical warfare are fascinating. Finding commonalities and dissimilarities between various groups reveals how diverse the natural world can be.

Poisonous organisms differ from venomous ones in that the noxious chemical is not injected but can be injurious if eaten or touched or inhaled (e.g., smoke from burning poison ivy). Some toxins can enter the bloodstream through a cut or the lining of mucus membranes. Poison ivy produces an oily substance that causes dermatitis in some people upon contact with leaves, stems, or roots. Death angel mushrooms and poison hemlock produce chemicals that are harmful if eaten. Common

Q: My friend says that some frogs are poisonous. However, he also says that toads do not cause warts. What is the truth?

A: Your friend is correct on both counts. Toads have bumpy skin and some of these bumps are glands that produce toxins, but scientists have no evidence that a person can get warts from touching a toad, despite the common superstition. And although many, perhaps most, frogs and toads have at least some toxins produced by glands in their skin, no frogs inject venom through fangs or stingers. Hence frogs are poisonous but not venomous.

garden toads secrete distasteful toxins from skin glands.

The primary function of injection by many venomous animals is to acquire their prey, although toxic chemicals are also used defensively. Examples of venomous injection methods include the nematocysts of a jellyfish, the serrated tail barb (a.k.a. stinger) of a stingray, and the stinging hairs of some caterpillars. Even some plants, such as the common stinging nettle, are technically venomous (rather than poisonous) as tiny hairs on the leaves and stem can penetrate the skin and release histamines when someone brushes against them. My grandchildren are no longer a threat to stinging nettles because all have learned firsthand the perils of tramping through a patch of this well-armed plant.

A few of the world's mammals qualify as venomous, the best known being the male duckbill platypus of Australia. A sharp spur on each hind foot is connected to a venom gland and a duct that transfers the toxin to the barbed structure. Short-tailed shrews of the eastern United States have toxic saliva that enters the body of prey, or would-be predators, when the shrew bites it.

No venomous birds have yet been discovered and presumably none exist. But the pitohui birds in New Guinea have poisonous skin and feathers. The most toxic of the New Guinea birds is the hooded pitohui, a small, foul-smelling, orange-and-black creature with a crest like a tufted titmouse. While collecting and preparing the first specimens of hooded pitohui birds, the investigators suffered from bouts of sneezing, along with numbness and burning of the mouth and nasal lining. As is often the case with scientific discoveries, the local populace already knew about the phenomenon. A 1977 book on folklore of the central Highlands Region in Papua, New Guinea, mentions that local residents said the skin of the hooded pitohui "is bitter and puckers the mouth." They referred to it as a "rubbish bird" and advised that it not be eaten "unless it was skinned and specially prepared." I have not sought out the recipe.

The chemical composition of the pitohui bird poison is similar to that in the skin of poison dart frogs of Colombia, South America. These deadly little frogs secrete a toxic chemical, a type of alkaloid that makes them unpalatable to other animals. If eaten or injected, as with the tip of a man-made dart used by Colombian natives to hunt prey, the toxic material has an immediate effect on the nervous system. The chemical is a powerful deterrent, and predators avoid poison frogs as a source of prey. Presumably the poison operates in a similar fashion for the New Guinea pitohui birds by discouraging typical predators such as snakes, other birds, and mammals from having an otherwise tasty meal.

Any increase in our knowledge of the natural world is of value if it raises our intellectual consciousness. Such discoveries help us to better understand the differences and similarities of various organisms as well as to appreciate the many ways nature has of solving problems. Knowing that natural poisons (including injected venoms) are an everyday part of nature makes it all the more interesting—although I understand how a bad case of poison ivy might dampen one's enthusiasm for the great outdoors.

Few Folks Get to See a Glossy Strangle a Craw

I received the request via email: "Let me know when you see that glossy strangle a craw." I replied that I would.

I have not become a spy and this was not code for some hush-hush operation designed to exterminate a criminal mastermind. It was a request from Noel Vick, who had given me the glossy crayfish snake, to let him know if I was able to observe the snake preparing to eat its favorite prey, crawfish—or as some call them, crayfish. I was able to catch several craws that did get strangled, but I never saw that glossy in action.

These small, harmless watersnakes have a black, brown, or olive-green body that has an iridescent sheen when wet. When a glossy crayfish snake captures its favorite prey, it wraps coils around the pinching claws to keep them out of play and proceeds to swallow this dangerous catch backward, tail first. An impressive feat. The snakes are only found where crawfish abound, so their presence is usually a sign of a healthy aquatic environment.

Glossy crayfish snakes deliver several ecological messages. The first is that they are part of our hidden biodiversity. Though seldom seen, they are nonetheless susceptible to pollution and other habitat degradations that adversely affect crayfish and other members of the natural food web. We tend to focus on highly visible species without considering that our actions can also eliminate those we do not see.

Another point is that capture records of glossy crayfish snakes are too scarce to determine their conservation status. Consequently, they are offered no environmental protection because their ecology is too poorly understood for scientists to advise what should be done or not done to protect them. Their geographic range includes a spotty distribution in the coastal plain from Virginia to Texas. Most herpetologists have never seen one in the wild. An out-of-sight out-of-mind mentality toward conservation is not in the best interest of countless unseen species of native wildlife that are affected by human actions. But to protect them, we must first understand their ecology.

A third lesson is that under the right circumstances a single persistent scientist can advance ecological

Q: My high school biology class is discussing environmental topics of current interest, including pythons in the Florida Everglades. Have pythons really invaded Florida? If so, to what extent?

Someone said you wrote the foreword to a book on the subject. Can you tell me the name of the book?

A: The book is *Invasive Pythons in the United States: Ecology of an Introduced Predator* (University of Georgia Press, 2011). Written by Michael E. Dorcas and John D. Willson, it offers a thorough assessment of the status of pythons in the Everglades. The python introductions in southern Florida, especially in the Everglades, are unequivocally a problem and have been well studied. Many scientists are working to understand the biology of Burmese and other python species and determine what can be done to control their spread. Their negative environmental impact is staggering because native wildlife in Florida (including deer, herons, and raccoons) are not adapted to coexist with giant constrictors and readily fall prey to the big predators.

knowledge. For example, Phil Vogrinc, a University of Arkansas student who conducts his thesis research in South Carolina wetlands, has probably caught more glossy crayfish snakes than anyone else in the world. In two summers he captured more than 800 snakes, of which 154 were glossy crayfish snakes. He has gathered more information about the ecology and behavior of the species than herpetologists have done collectively since the snake was first discovered almost two hundred years ago.

Such ecological research is not easy. Persistence, perseverance, and hard work are essential. Phil's study involved using "minnow traps," which are plastic mesh funnels about a foot in length with an opening at either end. Traps are set with a few inches protruding out of the water so that a snake that swims in can still breathe. For days at a time he set out hundreds of minnow traps in wetlands. Walking waist deep through mucky habitats choked with aquatic vegetation several hours a day, he checked each trap at least once daily; twice if it rained, to ensure traps remained above water. Phil's major professor, J. D. Willson, used the same technique extensively while studying wetland snakes in the same area during his dissertation research with the University of Georgia a few years earlier.

How many animals out there that are part of our hidden biodiversity are waiting for someone to develop the right technique to study them—and then to persevere in gathering information about them? Only when we understand the ecology of a species can we determine what environmental protection it needs. Fortunately for the glossy crayfish snake, Phil Vogrinc's research promises to be enlightening about this secretive species.

Phil was able to videotape the first documentation that glossy crayfish snakes actually do use coils to constrict around the pinching claws of their favorite prey before they eat them. He sent me a copy of the video so I was able to email Noel Vick to let him know that I had indeed seen a glossy strangle a craw.

I Wish Everyone Could Visit
Glacier Bay

Old ecologists are wont to tell young ecologists that to fully appreciate the earth's myriad mutualistic and competitive plant and animal interactions they must visit the tropics. Seldom does anyone suggest visiting the other end of the environmental spectrum—the frozen northland—for comparative purposes. For anyone wanting to accomplish that goal, I recommend a trip to Glacier Bay National Park (GBNP) in southeastern Alaska.

GBNP, which encompasses 5,150 square miles, is one of the largest national parks. It is larger than the states of Rhode Island and Delaware combined. About twenty thousand years ago, following the end of the last ice age, the region warmed up. The phenomenon of retreating ice revealed open rocky terrain that became habitable by wildlife, a consequence of microscale climate change in a formerly unvegetated habitat. During the ensuing, relatively warmer, times, the region became biologically productive and humans, the Tlingit Indians, inhabited the region. Then came the so-called Little Ice Age. Giant glaciers formed in Alaska and advanced down the valleys, forcing the Tlingits inland to warmer areas in the 1600s. As in earlier ice ages, the region was covered with thousands of square miles of snow, glaciers, and other forms of ice—and no open water.

Capt. George Vancouver of England sailed to the region in 1794. His ship's log did not mention Glacier Bay as a majestic place to visit. Simple explanation. At the time, 215 years ago, the bay was not majestic. It was barely a bay. Instead of an expanse of water, Vancouver found a hunk of ice known as the Grand Pacific Glacier, solid-packed ice roughly twenty miles wide and three-fourths of a mile thick. In 1879, twelve years after Seward's Folly had been purchased from Russia, came another visitor, John Muir. He traveled by water forty miles farther up Glacier Bay than Vancouver had. In less than a century, a solid, impenetrable glacier had retreated almost fifty miles, leaving a scoured valley filled with navigable sea-water. Today, the face of the glacier has retreated another twenty-five miles, creating a wide, sixty-five-mile-long bay with numerous fjords and more than a dozen side glaciers, most of which retreat farther every year.

In 1925 Pres. Calvin Coolidge designated Glacier Bay and surrounding areas as a national monument. Con-

Q: Where do turtles go when it's cold? I just read an article about someone in Florida who brought their pet baby turtles inside and put them in the bathtub to protect them during a cold snap. The story focused on the pet golden retriever that ate one of the turtles and was taken to the vet. I don't actually care what happens to turtles eaten by dogs but want to know if cold is really a problem for turtles in the wild—freshwater turtles, sea turtles, tortoises?

A: Cold is not a "problem" under natural conditions for native animals or plants. Individuals unable to tolerate seasonal changes in temperature under natural conditions are typically eliminated through natural selection. Thus, the genetic makeup of future populations is determined by the survivors, individuals who were able to withstand cold. Ancestors of all turtles on earth today survived whatever natural conditions of cold they were confronted with in their natural environment. None are likely to evolve to deal with being put in a bathtub where a dog can eat them.

gress made it a national park in 1980, and the United Nations recognized the area as a world heritage site in 1992. This vast and impressive region deserves all the environmental recognition and protection we can give it. The wildlife is astounding—on land, in the sea, and in the air. Seals, humpback whales, and sea otters abound. Puffins, bald eagles, and glaucous-winged gulls add another level of environmental magnificence. And bears, moose, and mountain goats inhabit a land often covered by magnificent virgin spruce forests.

Environmental views concerning Alaska and its natural resources run the gamut. Some people see a land where rugged independence translates into unrestrained oil exploration in the Arctic National Wildlife Reserve and clear-cutting in old-growth national forests. Others feel just as strongly that unspoiled habitats and environmental sustainability should be the order of the day. Some of the debates will not be resolved for decades.

One point, at least, is irrefutable: Glacier Bay offers a fascinating environmental lesson in ecosystem development. Starting at the cold end, near the face of the glacier, only mosses and lichens cover the exposed rocky surfaces. Along the many cascading streams of melting ice that flow into the bay, willows develop after a few years. Then come alder and cottonwoods, followed eventually by hemlocks and spruce forests. Although any plant might appear at any time, the biological succession is a relatively orderly process from lichens to mature spruce forests. Soon after plants become established, the terrestrial mammals appear—moose, mountain goats, bears, and wolves. Birds and sea mammals arrive when the waterways open up, providing an abundance of fish. The ecosystem is elegant in its simplicity and offers many ecological lessons.

I concur with the common wisdom that young ecologists should visit wild tropical ecosystems with their intricate evolutionary relationships created over eons. Indeed anyone interested in nature would benefit from such a visit. But I also recommend a visit to Glacier Bay, which offers a simple yet dramatic view of how ecosystems are born.

Sea Otters Are Unique

Sea otters are on my top ten list of "most appealing mammals," along with pandas, beluga whales, meerkats, and, of course, puppies. But sea otters are not just cute, they are also one of the most resilient animals in the world. Which is a good thing, because in addition to living in a harsh habitat, they were once hunted nearly to extinction for their fur.

These engaging creatures come as close as most animals get to qualifying for the label "unique." Being unique is not easy because you have to be the only one in a category. Sea otters are not the only marine mammal, which includes dolphins, seals, and whales, but they are the smallest. They are not the only animal in the weasel family, which includes badgers, skunks, and wolverines, but they are the only ones without the scent glands that produce the strong-smelling musk characteristic of the rest of the family.

In the one-of-a-kind category sea otters are the only marine mammal that has no blubber or other insulating layer of fat to help it survive in cold waters. Instead, they have the densest fur among all mammals (more than a million hairs per square inch!), which traps air that provides insulation. The density of hair among humans is typically less than three thousand hairs per square inch. Sea otters can weigh up to a hundred pounds, making them one of the heaviest members of the weasel family, but they float with ease because of air trapped in the fur.

Their diet includes fish, abalones and other mollusks, sea urchins, crabs, and other sea creatures, most of which they dive to the sea floor to obtain. They have lungs that are more than twice the size of most comparably sized mammals. They have been known to dive as deep as three hundred feet and commonly stay under water two to four minutes. Among the myriad traits that set them apart from other animals is their use of tools. They use rocks to dislodge abalones, which attach themselves to hard surfaces and must be pried loose. Furthermore, a sea otter, which typically floats on its back while eating, will set a rock on its belly and open the abalone shell by pounding it against the rock.

I rubbed my hand over a sea otter pelt once; it was the softest fur I have ever felt. Their fur was the reason for their near demise, initially at the hands of the Russians in the 1700s. In fact, the primary reason for the Russian occupation of Alaska before we bought it for a few cents an acre was the sea otter fur trade. At the time, sea otter pelts were considered the most valuable in the world.

The Russians recruited people native to the Aleutian Islands for the fur trade beginning in 1741, when the

Q: I saw a video of a cat on a boat playing with a dolphin. Both animals seemed to be enjoying themselves. The dolphin even looked like it was smiling. Was this real or some kind of filming trickery?

A: Apparently you have never had a dolphin or a cat for a pet. A dolphin smiles perpetually, not because it thinks everything it does is funny but because that it is the way its head and mouth are shaped. When cats are not sleeping or eating or annoying their owner in some way, they are looking for something to play with—their shadow, a cardboard box, a dolphin. I found the video you're referring to online, and it is as real as it gets when dolphins and cats have nothing else to do. The dolphin even looks like it is laughing. They like being filmed.

number of sea otters worldwide was estimated to be as high as three hundred thousand. The unregulated killing of sea otters continued in most parts of the animal's geographic range from northern Japan, across the Aleutians, and down the western US coast to Baja California until 1911. In that year four nations—Russia, Japan, Great Britain, and the United States—established a treaty to protect the species. At the time fewer than two thousand sea otters were left in the world, and some estimates place the number as low as a thousand.

We came very close to wiping out this marvelous species through unsustainable harvesting. We actually finished the job on the East Coast with the sea mink, which became extinct in the 1800s as a result of overhunting for the fur trade. Sea otters are tough, but they take several years to mature, usually have only one pup, and can live more than twenty years. Taken together those factors mean the species cannot recover quickly from a population decline.

In the past hundred years the worldwide population has gradually increased. Although exact numbers are hard to come by, about a hundred thousand wild sea otters are estimated to be alive today. I am glad the sea otter is still around to be placed on my top ten list of most appealing mammals.

Controlled Access Works Best
for Some Parks

"Run from a moose. Stand your ground with a bear." So reads some of the information given to visitors to Denali National Park and Preserve in Alaska.

The rationale for the guidance is that a moose will not chase you very far, so you can outrun it. But a grizzly bear will, at better than thirty-five miles an hour. So your best option is to just stand your ground and identify yourself to a bear in a nonthreatening manner. Apparently the approach works, as no one has ever been killed by a grizzly in Denali.

William Henry Seward, US secretary of state from 1861 to 1869, was chastised for his "folly" when he purchased Alaska from Russia in 1867. His justification was primarily to acquire natural resources for the United States. He probably had in mind minerals and ores. But as we ended up with what was then called Mount McKinley (now Denali), the tallest mountain in North America, and with a national park bigger than the land area of Hawaii, we also acquired the natural resources of native habitats, plants, and animals.

Denali is different from our other national parks in many ways. For one thing, the park is bigger than Yellowstone, the Everglades, and the Great Smoky Mountains National Parks combined. The park surely protects the largest population of grizzly bears anywhere, along with a host of other wild animals, such as wolves, Dall sheep, caribou, and moose. All are allowed to interact with each other and their habitats the way they were meant to—with minimal human distraction.

This feature is the one I like most about Denali National Park—the land is for the animals and their habitats, with us humans as noninvasive observers only. Private vehicles can only travel on the first fourteen miles of the eighty-nine-mile road, mostly gravel, that traverses the park. To go into the interior, you must hike or take a shuttle bus. Backcountry camping requires a permit. No fires or guns are allowed, nor is subsistence hunting or fishing. Lewis and Clark would not have liked those restrictions but, sadly, those days are over.

The shuttle bus idea is a great one and should probably be implemented at some of our lower-forty-eight national parks. Animals can be viewed and photographed but only by controlled access. Some people may maintain that to have a full wildlife experience, you need to be out there in the river bottoms, on the mountains, and with the animals. You can do that at Denali, if you like,

Q: Is it safe and also legal to feed wild animals? I know people who feed squirrels and pigeons in parks, have bird feeders in their backyards, or throw food to chipmunks around picnic tables at a state park. But what about deer or bears? Are there any restrictions on feeding these or other animals?

A: The answer to your general question about whether it is advisable or legal to feed wildlife all depends on the species. It is against the law to feed wild dolphins, alligators, and species on the federal endangered species list or in national parks. On the other hand, feeding birds, squirrels, and chipmunks is viewed as acceptable by most people under most circumstances; lots of people do so, usually without adversely affecting themselves or the animals. As for deer, which are game species, the rules may vary from state to state or even within a state. Quite aside from any legal aspects, feeding a black bear with sharp teeth and big claws is not a good idea, no matter how cute and cuddly it appears to be.

by getting off one shuttle, walking any direction, and later hitching a ride on another shuttle.

Limiting private vehicle traffic into our sizable national parks is really the closest we can come to achieving true protection of the sanctity of a natural preserve while still allowing it to be visited. Tour guides who drive the shuttles keep a slow pace and are trained to sight wildlife, thus avoiding animal-car collisions.

Denali successfully addresses the conflict between keeping wilderness wild, that is, sans human intervention, and allowing visitors to enjoy the sights, sounds, and smells of nature. You can experience it all, you just can't alter it, intentionally or inadvertently. If we want to keep some of the country wild, we must accept the idea of limited access.

Our national parks are beginning to reach visitor capacity each year, which suggests we have a citizenry who really appreciate nature and wildlife. As the human population continues to increase—and more people mean more potential visitors to nature preserves—the only solution for fulfilling our craving for visiting natural habitats is to protect more of them.

Do you suppose our congressional leaders would think it folly to support a movement to acquire more lands throughout the country and protect our natural habitats through controlled access? Immediately coming to mind are the millions of acres of lands protected environmentally by the Departments of Defense and Energy. These already have limited access and would not entail buying private lands.

Setting aside federal land that is now idle would be an ecologically sound way to conserve our natural resources for posterity across the country. Would Congress run from such an idea as if it were a moose or be stalwart and stand their ground as if it were a grizzly?

The Badlands Offer an
Environmental Paradox

The bare hills, catching the gold of the morning sun, looked like a distant painting rather than real life. Obvious wildlife was absent on the barren hillsides, which were devoid of vegetation save a scattering of small bushes and occasional trees, probably fewer than ten per acre. The region seemed perfect for a case study about the many faces of erosion. Yet the scene was starkly beautiful, especially when the exposed soils revealed their colors at sunrise and sunset.

The striking panorama brought to mind a paradox about environmental attitudes, because I could have been viewing either of two habitats: one natural, the other man-made. Suppose the Badlands where I stood were a consequence of human development and degradation of forest lands or prairie rather than the result of millions of years of natural processes.

Looking over the Badlands brought back memories from my visit years before to a similar-looking habitat—Copper Hill, Tennessee. Fumes from copper smelting plants had destroyed all vegetation from hundreds of acres, creating textbook example of environmental disaster. Ironically, the environmental landscape of both the Badlands and Copper Hill have many similarities, yet one is considered magnificent; the other, an environmental embarrassment. Suppose the circumstances were reversed, with naturally caused fumes, perhaps from some volcanic activity, having created the view at Copper Hill and an ill-planned agricultural project having been responsible for the Badlands. Would attitudes about the two places also be reversed? Would Copper Hill be a

Q: I am a pastor in Worland, Wyoming, working on a sermon on God's call to Moses in Exodus 4. I'm curious to know about the dangers of picking a venomous snake up by the tail.

A: Picking up a venomous snake by the tail or any other way always involves an element of risk. Knowing whether a person who gives a snake an opportunity to bite yet goes unbitten is a product of divine providence, good luck, or a disinterested snake is difficult to ascertain. One sure way to avoid the problem is not to pick up the snake. I can't think of a more appropriate time for using the phrase "the Lord helps those who help themselves" because most snakebites in the United States are a result of people picking up the snake.

national monument; the Badlands, an indictment of poor environmental planning?

Another ironic environmental situation involves clear-cutting, the timber industry's practice of removing all the trees and other vegetation in an area before planting crop trees. I once was leading a tour group from a national conservation society through a southeastern forest habitat, a public land area managed by the US Forest Service. This particular group was rather outspoken on their opinion that clear-cutting was a shameful practice and should not be permitted.

As we drove through one heavily forested section, we came to an area where hundreds of acres had been denuded of all trees. Several members of the group began to complain about such a practice being allowed on public lands. Some wanted to take pictures to document the devastation that could be caused by clear-cutting, an obvious example of forestry mismanagement. I let them rant for a bit before revealing the truth about the treeless expanse: the entire section of forest had been leveled by a tornado the previous spring. Once the group realized that a natural event, not human interference, had caused the devastation, the area became environmentally acceptable. Something to marvel at, not condemn.

Why are such different opinions held about the Badlands and Copper Hill when virtually indistinguishable close-up photographs of the two sites could be taken under the right conditions? Why do people have different attitudes about the hot waters of Yellowstone compared to those released as cooling waters from reactors? One is a natural wonder; the other is viewed as pollution. Contrasts in environmental attitudes clearly result from events that go beyond the visual appearance of a habitat or the wildlife experiences one might have in it. We should ask ourselves why it matters to us what caused a particular environmental situation.

The answer to the irony must go beyond the simplistic one of our having an inborn love of natural things. Do we harbor an inherent resentment, apprehension, and distrust when other humans control our natural resources? Do we accept nature's rearrangement of the world's environments but become uneasy about modifications that demonstrate someone else's ability to control our environmental welfare? Perhaps these deeper, unexplored feelings are the underlying cause of many of our environmental conflicts.

Maine Has Abundant Opportunities for Nature Watching

I enjoy hot, humid, oppressive summer weather as much as the next person, so when I was invited to spend a few days last week with a friend in Waldo County, Maine, I quickly said yes. I had dug out my favorite fleece jacket before I even accepted the offer.

We went by ferry to Islesboro, an island on Penobscot Bay, halfway up the Maine coast. My host was J. D. Willson, an ecologist at the Savannah River Ecology Lab; he frequently finds he has work to do in Maine during the summer.

Though I'm not an ardent birder, I enjoy nature watching in general. During my visit to Maine I could have added several bird species to my life list, if I kept a life list. On a boat trip to an island called Matinicus Rock, known for its variety of seabirds, we saw common and Arctic terns, dainty Wilson's storm-petrels, and four species of auks. For me, the truly exciting auk was the puffin. These cute, white-faced, black-capped birds with the fat, red-tipped bill looked just like the pictures we have all seen of them sitting on rocks. The ones we saw were also swimming, flying, and diving.

J. D. is an expert on North American birds and was able to identify every one we saw or heard on land and at sea. Our most dramatic bird encounter involved three species and was a true circle-of-life adventure. One foggy day, we were approaching one of the hundreds of gigantic rocks that peek above the surface at high tide and must be skirted by lobstermen in their boats and by us in ours. Peering forward, J. D. said, "Must be an eagle ahead, the way those sea gulls are acting." I peered into the gloom and could barely make out the little island, let alone any birds. I scrunched into my fleece jacket and, being polite, said, "Oh, yeah."

As the fog cleared, I saw what he meant. About a hundred herring gulls were swarming around the rocky island like bees around a hive. A short distance from the island a pair of bald eagles were circling, each being chased by a sea gull the way a mockingbird chases a crow or a crow chases a hawk. Then J. D. pointed out a bird we had not yet seen on this trip—an eider duck. One of these ducks was of special interest. It was a young one two hundred yards offshore from the little island and it was being circled by one of the eagles. While we watched, the eagle made an unsuccessful dive for the duck, which had gone under water.

The drama played out as the little duck popped to the surface; the gull following the eagle abruptly did a barrel roll and plucked the baby duck from the water. The gull headed toward the island with the peeping duck in its mouth. The eagle, meanwhile, was leaving the scene when it realized the gull was no longer pestering it. Not only that, the gull had a meal in its mouth! J. D. and I watched in wonder as the eagle did a 180 in midflight then turned on the afterburners. It flew within a few feet of our boat in pursuit of the duck-snatching gull and completely ignored our presence.

With its overpowering speed, the eagle caught up with the gull near the edge of the island. In cartoon fashion, the gull raised both wings in surrender and opened its mouth. The eagle swooped up the eider duck in midair and left the scene with the baby duck in its beak. As the final act came to a close, we looked over at a harbor seal with its head poking out of the water. It was swimming in the direction of another eider duck bobbing in the water. Little eider ducks are apparently a popular menu item in the cold waters of the northern Atlantic.

When I got back to South Carolina, I made a point of telling J. D. that I would keep my fleece jacket close at hand in case we needed to check on the ecology of the seals, puffins, eagles, gulls, or eider ducks again that summer or the next.

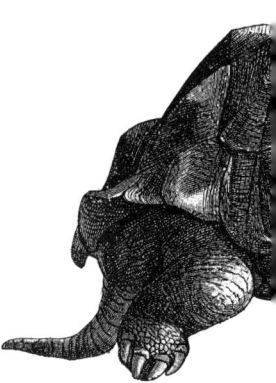

Autumn

Take a Walk in a Southern Stream

When I saw the bullfrog sitting up in the bush with all the flowers, I knew the trip had been worthwhile. That's the great thing about getting out in the woods in autumn before the cold spells hit. Animals and plants abound, and you never know what you might see.

I was with John Jensen, the herpetologist with Georgia's Department of Natural Resources, and Kurt Buhlmann, the turtle specialist with Conservation International. Since all of us were overdue for a field trip, we decided to take an excursion out to a coastal plain swamp in South Carolina. John wanted to catch a rainbow snake, so that became our mission. Searching for the rare and beautiful rainbow snake is always a good excuse for an outing. When we saw the bullfrog we were wading shin deep in a clear, woodland stream, a tributary into the swamp. We were not really lost, I explained, because I was sure this particular stream came out near a dam where it flowed over a spillway. An hour or so before, I had admitted that I was not sure how far upstream the dam was.

Nonetheless, trekking along in a crystal clear, sandy-bottomed stream on an almost-cool autumn day can be delightful and make you glad to be out of the office. Numerous plants bloom in the fall, and the banks of most streams in the South are adorned with shrubs and solitary plants with sweet-smelling flowers. We saw bright red cardinal flowers along the stream's swampy margins. Shrubs with perfumed cream-colored flowers were on the higher banks. Yellow composites of different sorts were in the drier areas. My glance toward one of the flowering shrubs brought the amazing sight of the bullfrog.

Q: This may sound like a dumb question, but a friend says she knew someone from California who smoked toads at a campfire on a beach. Could this be true? I think she is pulling my leg, so she said to ask you.

A: The first rule of thumb about tales from California is to believe anything you hear until someone proves it is false, especially if it happened on a beach. Your friend's account could certainly be true. Let me clarify one point: Toads are not smoked like oysters or toasted like marshmallows. Toad smoking, which evolved from toad licking, involves the use of cigarette papers.

In toad licking, one laps up toxic secretions from paratoid glands on a toad's head to achieve psychedelic effects. A significant drawback of this behavior is that a frequent toad licker can become ill, maybe even die. So clever Californians found a way around the problem. Because heat breaks down the toxic components in the toad's glandular secretions without affecting the sought-after hallucinogenic compounds, toad secretions can be dried and rolled in cigarette papers. Hey presto, a puff or two and you're ready for your trip.

But beware. Besides being a bizarre practice, smoking or licking a toad has legal implications. The hallucinogens extracted from toads are an illegal substance. I personally doubt that either toad licking or toad smoking will ever become a serious problem, but let me go on record as saying that ingesting poison may be hazardous to your health. So think before you lick or inhale.

This was the first time any of us, all experienced herpetologists, had ever seen a bullfrog sit high in a bush, fully five feet above the ground, but there it was. We all knew that just because we had never seen such a thing before did not mean it was not a common event. And being ecologists, we assumed there must be an explanation. Within minutes, what we decided was part of the explanation flew into view.

I wish I could say we actually saw the event we assume occurs when bullfrogs perch in bushes (something, you will remember, that none of us had ever heard of or seen). However, we can speculate based on circumstantial evidence. Let me show you how the science of ecology sometimes works.

As we looked at the enormous bullfrog (probably almost a foot long when stretched out) perched majestically in the top of the bush, Kurt pointed out a cardinal flower growing nearby. I commented that cardinal flowers are one of the last natural feeding sources for hummingbirds that migrate south during this season. As I said this, Kurt pointed above the bush: "And there one is." A hummingbird zipped into view, hovered for a moment to watch the three of us standing in the stream, and disappeared. John then noted that "maybe the bullfrog is sitting in wait for a hummingbird meal."

We did not see the bullfrog eat a hummingbird, but I have no doubt that it had climbed into the bush to sit amid a smorgasbord of bird and insect prey. I looked upstream to the next bush like that one and noticed it was covered with butterflies, as this bush had probably been before we three humans blundered onto the scene. The bullfrog was simply doing what most animals spend much of their time doing—looking for something to eat. And bullfrogs have been documented to eat everything from birds to small snakes to baby alligators.

We never did see a rainbow snake on that excursion. But sighting the bullfrog was its own reward. And it reminded us that the unusual and unexpected may be encountered in nature at any time. You don't have to get lost in a stream; you just have to walk outside and look around.

Are Large Black Cats More Than a Halloween Phenomenon?

Black cats and Halloween are a natural combination. In fact, some superstitious folk think they come out in greater numbers on Halloween night, maybe riding on the back of a witch's broomstick or slinking across someone's path to bring bad luck. From an ecological standpoint, black cats have an element of mystery that has nothing to do with superstition. Over the last forty years I have had no fewer than a dozen people tell me that they have seen a black panther. These are not people who have heard other people say they saw one but instead are convinced that they encountered a large black cat themselves. We are not talking about in zoos, books, or movies but in the wild at various places in the southern United States. Are the stories true? Do giant black cats really exist in North America?

The black panther of jungle lore certainly exists. Black phases of leopards in Asia and Africa and of jaguars in tropical America occur in the wild. The genetic phenomenon known as melanism, which results in an individual being almost completely black, occurs in many mammal species in which shades of white, gray, tan, or brown are typically more prevalent. Melanism has been documented in coyotes, gray squirrels, and even rarely in white-tailed deer, producing almost pure black individuals.

In rare instances a native US species of large cat, the bobcat, can be completely black. Bobcats occur throughout most of the United States, southern Canada, and Mexico. Photographs and museum skins offer scientific verification that such individuals exist. Most of the verified records are from Florida, but at least one black bobcat was found in eastern Canada. Presumably, the genetic condition that causes a bobcat to be black could occur geographically anywhere in between Florida and Canada or maybe anywhere within the range of the bobcat.

When people report seeing a "black panther" in the wild in North America, they usually mean the mountain lion (a.k.a. cougar, puma, catamount). And reports of people seeing black mountain lions abound. Yet, no photograph, no carcass, no scientific proof of any sort has ever been provided to indicate that a mountain lion can be solid black. This does not mean that a melanistic mountain lion does not exist somewhere or did not exist in the past. It only means that its existence has not been verified.

I can give several plausible explanations for why someone might claim to have seen a black panther. The

Q: I would like information about a giant salamander from the Trinity Alps in California that is said to be from five to eight feet long. I have heard that they do indeed exist.

A: The largest known salamanders in North America are aquatic species that live in the eastern United States. The bulkiest is the hellbender of mountain streams, with a record length of two and a half feet. The greater siren of the Southeast reaches lengths of over three feet. The largest American salamander, the two-toed amphiuma, has a record length of almost four feet. The world's largest salamander, from Japan, is closely related to the hellbender and gets more than five feet long. As the Trinity Alps are in northern California, people may be referring to the Pacific giant salamander, the largest terrestrial salamander in the country. Some approach a foot in length. Any salamander much larger than that from the western states would be worth reporting to the nearest zoo.

first is that the person saw a black bobcat. Mistaking a bobcat for a mountain lion may sound far-fetched because bobcats are smaller and have shorter tails, but I know this can happen. Three different times I have gone to look at a road-killed cat in South Carolina that someone had declared was a dead mountain lion. All were large bobcats, even though the person who reported it as a mountain lion had actually looked at the carcass. I know a wildlife biologist in Tennessee who investigated the killing of a "mountain lion." When he arrived at the scene and examined the carcass, he had to break the news to the hunter that he had killed a forty-pound house cat! This one was yellowish-brown, but imagine if it had been solid black. How many days would this long-tailed cat have to roam a neighborhood before people were reporting a black panther?

Another explanation is that a person might mistake a large, long-tailed black dog or coyote for a big cat, particularly at night or in fading light. To the embarrassment of two different individuals I know of who made such sightings, the animals left footprints that were made into plaster casts for identification—dog feet both times.

Another possibility exists in some states where it is legal to keep large house pets, including tigers, leopards, or even jaguars if you can get one. The black varieties of leopards and jaguars are proportionately more common among captive animals than in the wild because of selective breeding. Having a big cat escape from a zoo or a personal holding facility is certainly not unheard of. A big cat owner might be reluctant to admit that an enormous, stealthy predator had been unleashed in the neighborhood. So such a sighting might not be validated by the pet owner's saying, "Hey, that's my missing black leopard that just ate your poodle."

Another unfortunate explanation is the intentionally erroneous report of a dangerous animal that didn't actually exist. According to Chad Arment in *Varmints: Mystery Carnivores of North America* (Coachwhip Publications, 2010), a series of newspaper articles in the 1950s reported that a black panther had escaped from a carnival in Los Angeles. The subsequent hysteria resulted in the varmint's being seen at various spots around town. Several policemen claimed to have seen the 140-pound beast loose in an alley the first night; numerous citizens fueled the fantasy by reporting sightings of their own. The news stories are entertaining and demonstrate how human obsession with mystery animals can become a self-fulfilling prophecy. People think animals are there, so they see them, or at least think they do. The final newspaper account tells how the carnival owner eventually admitted he had "concocted the whole story."

My preferred explanation for people who believe they have seen a large, long-tailed black cat in the wild is that they actually did see a black mountain lion. Maybe they exist but are so rare that the carcass of one has never been found. Halloween is probably not the best time to be believed if you report such a sighting, but if you see one, it just might be real.

This Worm Is Creepier Than
a Halloween Ghoul

With Halloween on the horizon, I am reminded of a "mystery animal" someone once sent me. The enigmatic creature was described like this in an email: "A young boy discovered a specimen that our neighborhood is unfamiliar with. It is about twenty-four inches long, has the diameter of a toothpick, is brown in color, and has a flat nose and forked tail. It was found in a dirty puddle after we had excessive rains. We live in a South Carolina farming community. Do you have any idea what this might be?"

I asked the questioner if she could mail the specimen to me in a jar with tiny air holes in the top and damp paper towels inside. When I arrived at work two days later, my coworker Sarah Collie sat at her desk with a jar in her hands and a bemused look on her face. "Your worm has arrived," she announced.

And indeed, there it was, exactly as described—two feet of slender worm that looked like a tangled coil of thin copper wire with a forked tail. I had never seen a live one before and neither had any of the ecologists I checked with, although we all knew about such animals. After gathering the facts, I emailed my new pen pal that the animal was a horsehair worm, belonging to a poorly understood group, the hairworms, with more than three hundred species. Hairworms are in their own separate phylum whose biological relationship to other worms is ambiguous. The scientific name of one hairworm is *Gordium*, after the Gordian knot, which could not be untied (a difficulty Alexander the Great reportedly overcame by slicing the knot in half with his sword). After seeing the twisted knotty mess that a two-foot-long worm the diameter of a piano wire could get itself in, I thought the name most appropriate. The forked tail on this one meant it was a male.

Adults, which can reach a length of three feet, are free-living, meaning they are not parasitic. But the larvae are. Paradoxically, hairworms are special because they are so unspecialized. They have no digestive system, no respiratory system, and no circulatory system. Adult horsehair worms do not eat. After leaving the insects they grew up in, the males and females mate and reproduce in water. The female lays eggs that float in the water. If an egg is eaten by an insect, the egg hatches and the tiny parasite larva drills its way out of the insect's intestine and takes up residence in the body cavity. It feeds on the inside of the insect until it grows into a long worm ready to start the cycle again.

Invertebrates, including spiders and insects, serve as hosts for hairworm parasites. In what might appear to be the inspiration for the movie *Alien,* the larvae grow to the size of giant worms inside the unfortunate invertebrate host before they emerge. Imagine a three-foot-long worm inside a grasshopper! Hairworm parasites have little use

Q: Is it true that in most spider species the females are larger than the males? Do all female spiders eat the male after mating?

A: Female spiders are usually larger than males. For example, female golden silk spiders (*Nephila clavipes*) are about ten times larger than the males. Female spiders occasionally eat their mates. This is not always the case, but male spiders have extremely short life spans compared to females and most die after mating.

for people, although rare infections of humans in China, Japan, and Canada have been reported in medical and parasitological scientific journals.

How does the worm know it will end up in water so it can mate? A remarkable study by several French scientists on hairworms that infect crickets and grasshoppers may offer a partial explanation. In laboratory studies that examined the brain of insects, the investigators found that the hairworm parasite actually alters the behavior of the insect by producing molecules that enter the insect's central nervous system. The exact mechanisms are unknown, but chemical alterations in the brain make the insect jump into water and drown. Such abnormal insect behavior puts the now-developed worm where it wants to be. Effective—and lethal—brainwashing.

The final email from the woman with the horsehair worm displayed the kind of enthusiasm about the natural world that I like to see in people. "Thank you for this information. I learn something new every day!" That day, so did I.

Halloween Is a Time for Scary Thoughts

My daughter mentioned that one of the costumed characters running around her house getting ready for Halloween was Dracula, which made her think about the count's natural counterparts, vampire bats. She asked me if vampire bats actually occur in the region known as Transylvania and if they are found in the United States. My grandson wanted to know if I had ever seen a vampire bat.

Along with ghosts, witches, and gun-slinging cowboys, Dracula is a traditional costume choice for trick-or-treaters. As everyone knows, the nocturnal Count Dracula bites his victims on the neck, extracting blood from them, and, not coincidentally, turning them into vampires in the process. Bram Stoker's 1897 novel was modeled in part on tales dating back at least to the 1700s.

Although vampire bats are not native to the Old World and were unknown to science until the early 1800s, early explorers to the American tropics undoubtedly encountered them and probably offered anecdotal accounts when they returned home. Vampire bats, which occur from Mexico to South America, have been documented within two hundred miles of the United States. They are the only blood-sucking mammals in the world, except of course for Count Dracula and others of his ilk. Like Dracula, vampire bats come out at night and have sharp teeth with which to pierce the flesh of their prey; a vampire bat's saliva has an anticoagulant that increases blood flow.

Any mammal in a vampire bat's vicinity is fair game, but the most common prey are now cows, horses, and humans. A bat lands as inconspicuously as possible on a sleeping animal and administers a bite that may actually go unnoticed by the victim. Taking a snooze outdoors in vampire bat territory could be hazardous to your health, especially if you have consumed too much alcohol, which could lead to a deep sleep. A concern for someone bitten by a vampire bat, aside from the disturbing idea that a blood-sucking mammal has invaded your personal space, is that some carry rabies.

My own experience with vampire bats occurred in Costa Rica as I walked through a banana plantation one night looking for frogs and snakes. As I scanned the ground and trees with my flashlight, turning over dead banana leaves with a snake stick, I had the sensation of being watched. My snake-hunting companion was several yards ahead, so I turned, expecting to see someone following me. A bat was hovering effortlessly three feet

Q: How many kinds of bats live in the United States? Do they only come out at night? I am sure I saw one last week before dark.

A: More than thirty-five different species of bats are found in the United States if several tropical species that enter some of the southwestern states are included. A half dozen US species are widespread, ranging from coast to coast and from Canada to the Gulf States. About half a dozen species have geographic ranges that are restricted to the United States, but the vast majority are also found in Canada, Mexico, or both. Bats are able to be completely nocturnal by using supersonic sound production that allows them to find insect prey and avoid physical structures with echolocation. However, bats have tiny eyes and can see in the daytime; they are occasionally active an hour or so before dark and after sunrise.

away, just about neck high. It made no sound and, more disturbingly, did not leave. It just kept flapping its little wings and staring at me. Or at my carotid artery. I did not see it lick its chops, or whatever it is hungry bats lick, but it was clearly looking at me. (I may have imagined that it was sizing me up and thinking just how tasty I would be.)

I had read *Dracula* and seen several vampire movies, so I naturally assumed that the man in a black tuxedo had just turned into a bat and now planned to turn me into the living dead. Just kidding. Mostly. I knew that vampire bats were real and that concern for my welfare was no part of this bat's plan. So I took a swing at it with my snake stick. It flitted away, disappearing into the darkness. Did I hear it laugh at my feeble attempt to hit one out of the park?

I began walking again. When I looked back, it had returned. Presumably it was waiting for me to lie down and sleep so it could dine. Walking through a pitch-dark jungle can stir the imagination and mine was already plenty fertile. I caught up with my colleague and convinced him that no frogs or snakes were out that night so it was time to head back to camp.

Vampires, bats, witches, and ghosts are all part of the Halloween tradition. Suggestions have been made that because of global climate change, vampire bats may soon move from Mexico into the United States, with Texas and Florida being the most likely candidates for places to settle. Now that's a scary thought.

Africa Doesn't Need Bigfoot

When I left for Africa, stories in the US media were reporting Bigfoot sightings in Oregon, Alaska, and even Connecticut. Among the questions I have about Bigfoot is why one never gets accidentally shot during hunting season the way hunters and dogs do. Surely Bigfoot doesn't wear an orange hunting vest.

Another Bigfoot question: Why have such legends not reached Africa? I now know the answer to that one. Africa needs no Bigfoot myths because it already has plenty of the real thing—very large animals with very big feet. I saw two African wild dogs drinking water out of a rhinoceros footprint. Bigfoot indeed.

While visiting South Africa with Mike Dorcas, I heard no stories about the sort of creatures that reside in people's imagination rather than in the real world. Stories about legendary animals may exist in some parts of

Q: Should the public worry about the pythons in the Everglades? Are they a threat to humans?

A: The largest pythons are females, which can be more than twenty feet long, and could easily kill an adult human by constriction. One Florida python known to have eaten an eighty-pound deer could certainly swallow a small person. Pythons typically eat mammals, birds, and even alligators. A large predatory snake is unlikely to have any regard for whether it is eating a human or some other animal it considers prey. I'm not sure that it's useful to "worry" about being eaten by a python but being cautious in areas where they occur seems prudent.

Africa, but they are certainly not needed where we were. At least seven species of large mammals that roam wild have been documented as killing people without a hint of remorse. When you throw Africa's large vipers, mambas, and cobras into the mix, the I-can-kill-you-if-I-so-choose list passes a dozen.

We visited the world-famous Kruger National Park and surrounding South African nature preserves where predators and prey alike are protected by law. Upon entering Kruger, visitors are given strict instructions not to leave their vehicle. If you were to be fined for disregarding this rule, you would be considered lucky because that would mean you were still alive. The prohibition against leaving the vehicle is intended to protect visitors who might otherwise be trampled by massive hooves, gored by something the size of a Volkswagen bus, or eaten by a beast with claws and teeth that would rival the Iron Chef's knife collection.

Among the memorable animals we saw that helped us understand why it would be risky to leave the vehicle were two predators (lions and leopards) and three herbivores (African buffalos, rhinoceroses, and African elephants). These are known collectively as the big five. The term came into being back in the day when safari-goers were interested in shooting animals with a gun rather than a camera. Ernest Hemingway and other big game hunters of the 1930s considered the big five to be the most dangerous animals in Africa to hunt on foot, in part because they have little respect for human beings. Apparently they do not realize that people are special.

Any one of the big five can run twice as fast as the speediest Olympic runner, which means your odds of escaping on foot are exceedingly poor. A vehicle provides a substantial measure of safety. No one is quite sure why if people today stay inside their car, van, or bus they are usually safe. The assumption is that the animals view vehicles as sort of moving rocks. Usually animals simply ignore them and the people inside. I saw this many times as lions, hyenas, and leopards went about their business, apparently completely unconcerned that we were watching them from an open safari bus only ten feet away.

One animal not included in the big five, despite having the qualifying traits of enormous size, an ill-tempered nature, and the surprising speed of forty miles an hour, is the hippopotamus. Every safari guide we talked with confirmed that a person who ventured into the terrestrial area between the water and where a hippo was grazing should be sure his funeral arrangements had been made. This is not to say that hippos have a grudge against humans. Any animal that gets between them and the water is in imminent danger.

I was greatly impressed with South Africa. The conservation attitudes in the country are outstanding. Our safari guides were highly trained professionals. And the animals themselves provide the greatest wildlife show on earth. Meanwhile, if Bigfoot does exist and should happen to visit South Africa, I suggest that he stay inside the vehicle.

National Hunting and Fishing Day Is Good for the Environment

I'm not sure who first stated the following concept: people who hunt and fish are among the nation's foremost conservationists. I do know that Teddy Roosevelt was the first US president to strongly support the concept, and in 1972 Richard Nixon endorsed the idea with a signed proclamation. With the unanimous approval of Congress, Nixon designated the fourth Saturday of September as National Hunting and Fishing Day.

Some people do not accept the idea that killing animals in a forest or removing fish from a lake is good for wildlife. And for the individual animal that is shot or caught, the experience is clearly not beneficial. But here's the irony: hunters and anglers remain among the staunchest supporters of flourishing, thriving natural ecosystems. You cannot hunt game or catch sport fish if you do not have wild, undeveloped lands and clean, unpolluted waters. For the sportsperson who hunts or fishes, healthy habitats and clean environments are vitally important for maintaining sustainable use of game fish and animals.

The issue of animal rights often puts hunters and sometimes even anglers at odds with certain parts of society. Antihunting campaigns are rampant nationwide, and the conflicts are often bitter, sometimes political, and rarely if ever have an outcome that satisfies both sides. Somebody is going to be unhappy no matter what the outcome. Not surprisingly, one of the most vocal opponents of National Hunting and Fishing Day is PETA, the uncompromising animal rights organization. They petitioned President Obama in the first year of his presidency when he signed the presidential proclamation

Q: Organizations are working to capture pythons in Florida and remove them from parks and natural areas, and python-hunting contests have been held. Is hunting pythons an effective way to combat the invasion? What other steps should be taken at this time?

A: I do not think the python-hunting contests will be particularly effective as a control measure, though they may make people aware of the problem. Having scientific debates on how best to solve the problem would be a useful step, although not all scientists will agree on the best course to take. Python experts do not even agree on how far north the big snakes could extend their range. But scientific discussion in concert with continued research would clearly be an important step to take in developing strategies for coping with the invasive pythons.

designating NHF day, the way each president, Democrat or Republican, has done since Nixon.

The conflict between PETA and the hunting and fishing community will never be resolved to everyone's satisfaction. PETA and other animal rights activists have various reasons for launching antihunting campaigns that are directed toward particular species or areas, but they cannot assert that hunting is bad for the environment. Because hunters and anglers want to be assured that their target species are still around in the future, they support environmental protection.

To ensure that fish and wildlife populations are sustainable from one generation to the next, a diversity of natural habitats must be kept intact, unpolluted, and undisturbed. Hunters and anglers support these efforts with their attitudes about natural habitats and with their pocketbooks. Federal excise taxes on hunting equipment contribute directly to the support of land purchases, habitat protection, and wildlife management programs. Likewise, anglers throughout the country contribute significant amounts to public and private environmental protection enterprises by buying fishing licenses and equipment.

According to the US Fish and Wildlife Service's National Survey of Fishing, Hunting, and Wildlife-Associated Recreation in 2011, "Hunters, anglers, and wildlife watchers spent $145 billion on wildlife-related recreation." That's a highly significant contribution toward environmental protection, although it should be noted that the survey includes other wildlife-related recreation besides hunting and fishing. According to USFWS these other outdoor pursuits include "wildlife-watching activities such as observing, feeding, and photographing wildlife." Nonetheless, the hunting community, although a small proportion of the national population and getting smaller every decade, continues to have a major impact on environmental protection and preservation. Hunting revenues are substantial.

Habitat preservation is critical for all wildlife, not just game species. The major threat to most natural ecosystems and wildlife species today is habitat degradation and destruction. Irresponsible commercial development is a leading culprit when natural habitats are destroyed, then replaced with artificial ones where most native wildlife do poorly. In many states, hunting clubs preserve more natural habitat than most environmental organizations. The focus of such clubs may be on managing selected game species, but nongame wildlife also benefit. Some hunting clubs are exemplary models of private land ownership helping to preserve natural habitats.

So on the fourth Saturday of September, National Hunting and Fishing Day, if you know someone who hunts or fishes, consider taking a moment to say thanks for helping keep America wild.

What Are Our Top Ten
Environmental Problems?

Top ten lists are appealing, from top ten books to top ten YouTube videos to top ten news items you are not looking forward to reading this month, which for me would include political ads and hearing whether Justin Bieber is behaving himself. Getting young people to think about the environment is a worthwhile goal, and listing the top ten environmental problems could be a good classroom exercise for grammar school through high school.

The following list has changed little from a consolidation of views by my ecology graduate students at the University of Georgia in 2005. The original list included seven items that were somewhat interchangeable in regard to student opinions as to how they should be ranked. Items 1, 2, and 3 were the same in almost everyone's list, although not always in the same order.

My top ten environmental problems in order of increasing importance:

10. Air pollution. Uncontrolled releases by industry and the excessive use of fossil fuels have led to acid rain, dissolution of the ozone layer, smog, and the general elimination of "clean air."

9. Invasive plants and animals. The trouble resulting from fire ants across the southern United States, Burmese pythons in the Everglades, and many other regional environmental problems have a human origin related to the introduction of exotic species.

8. Global climate change. This has become such a volatile issue that debates between advocates and

disbelievers about potential impacts and solutions are seldom productive, with disputes even extending to the semantics of whether we should call it "global warming" or "global climate change." In any case, numerous credible scientific studies have documented changes in recent decades. A problem today, compared to the ancient past, is the impaired ability of some native species to respond appropriately to climate change as they have in past millennia because humans have compromised natural environments.

7. Pollution of marine habitats. The oceans are huge, but overharvesting and the degradation of marine environments are proceeding at a steady rate around the world, including a commercially extinct codfish industry and disappearing coral reefs along our coasts. The Deepwater Horizon oil spill of 2010 along the Gulf Coast, more popularly known as the BP oil disaster, comes to mind when the words "ocean" and "pollution" are mentioned.

6. Unsustainable agriculture. Humankind is depen-

Q: I am writing from Thessaloniki, Greece, to ask your opinion concerning the radiation of mobile antennas. How dangerous are they, and what are the proper protections that should be taken during the installation of such antennae in a populated area?

A: I checked with a colleague who is a radiation ecologist. He said he is unaware of any confirmed problem as far as radiation goes with such antennas, which are used in many parts of the world as transmission points for mobile phones. One of the greatest environmental hazards of any high antenna is that migrating birds that fly at night are killed in large numbers by colliding with the structure.

dent on food production, yet agricultural siltation, pesticide runoffs, and loss of natural habitats are constant threats to a healthy environment.

5. Threat of disease. Swine flu, West Nile virus, and mad cow disease are examples of ways we can be affected by unseen enemies. All are a consequence of human overcrowding, overconsumption, and invasive species.

4. Water quality and quantity. Sewage from cities, unregulated releases from industrial and agricultural sites, and dumping of wastes in the oceans collectively exacerbate the worldwide problem of water pollution. In many coastal regions, overuse of groundwater, which leads to saltwater intrusion, is a looming specter. Water wars are now a reality in the western states and even in the wetter Southeast, as evidenced by court cases involving Alabama, Georgia, and Florida.

3. Habitat loss, fragmentation, and degradation. The loss of natural habitats because of human development and deforestation is a major cause of the decline in biodiversity nationally and globally. Many species are on an inexorable path toward extinction because their native habitats are gone or despoiled.

2. Human overpopulation. Unchecked human population growth leads to overconsumption and associated world poverty. Many people would rank overpopulation as the number one cause of our other environmental problems. Virtually every problem above can be traced back to our simply having too many people for the resources available. Until political and religious leaders have the courage to address the issue of birth control on a global scale, most of our environmental problems will worsen before they get better.

1. Political apathy. I consider this the number one reason that the other nine environmental problems listed here are not being properly addressed. A clear indicator of this is that world leaders seldom acknowledge, let alone propose solutions to, environmental problems. For those leaders around the world chosen by a democratic election process, voters are not just condoning such apathy, they are participating in it.

What would schoolchildren consider to be the top ten environmental problems facing the United States and the world? Having a class offer their views should generate some interesting discussions. And one thing is certain, until we overcome our apathy, we will never overcome our other problems—environmental or otherwise.

Get a Head Start on This Year's
Science Fair

Next to parental guidance, most credit for today's environmental awareness among young people goes to schoolteachers. For much of the year they spend almost as many waking hours with children as their parents do. One of their teaching tools is science fairs.

The following two projects are ones that can be successfully completed by and educational for the student. I'm sure because I have suggested these projects before, and students have used them in science fairs. A gratifying moment was when a teacher at a school where I had given a talk introduced me to a sixth grader who had won their school science fair using one of the projects. I won't guarantee that either will win a science fair in the age of computer technology, but both relate to international environmental issues: the fossil fuel crisis and biodiversity.

The fossil fuel hypothesis

Shoppers driving into a mall would save both time and energy by selecting the closest readily available parking space when they enter the lot rather than driving around looking for one closer to the building or waiting for someone to pull out of a desirable spot.

The premise
Both gas and time are wasted by people driving around for a place near the store, while passing up open slots a few spaces farther away. The supposition is that

the odds favor getting into the store faster by taking the first available spot and walking a few extra feet, instead of searching for that prime spot. The same is true of waiting for someone to pull out of a spot closer to the store.

The procedure and questions

Stand in front of a mall or big grocery store during a busy time as cars come in. Use a stopwatch to time how long it takes drivers to park and how long it takes them to get to the front door. Do drivers who take a spot immediately, even though it is near the back, on average actually get to the front door faster, perhaps even while the searchers are still driving around? If it turns out to be true that those selecting the first spots reach the front door sooner, they not only save time but also use less gas and get more exercise. In a relatively short time a student can get several dozen data points, support or refute the hypothesis, and produce a science fair project that tells people something useful. Embellish the project by calculating how much extra gas the searchers use.

The biodiversity hypothesis

Plants and animals will live on any available space if given enough time, even on a vertical wall.

The premise

A 1981 book by Arnold Darlington titled *Ecology of Walls* asserts that walls comprise more than 10 percent of the area habitable by plants and animals in a city. Walls are all around us, providing habitats for many species. Included would be garden walls, the sides of houses and sheds, even the trunk of a big oak tree, which is just a natural wall.

The procedure and questions

Sample a variety of walls and record what lives on them. What variables affect the composition of species and the success of different kinds of organisms? Are walls with horizontal sections that create shelf space more likely to collect dirt and debris where seeds can root? Does compass direction matter for some species? For example, does moss grow mostly on the shady side of a wall? Do the wall's material and composition have a major influence on what clings to a particular wall? Does the age of the wall influence the vegetative character? What lives on the wall? Algae, moss, and lichens? Do vines and even small plants grow on crumbling walls? How about animals like lizards, treefrogs, spiders, millipedes, and a variety of insects?

Either project can be completed in two weeks and can be carried out in hot weather or cold. Consequently, students (and their parents) will have ample time to procrastinate until only a few days before they are due and still get the project completed. But don't wait. Imagine the data a student could accumulate to make the point that driving around looking for parking places is wasteful or to demonstrate that walls are important to the biodiversity of an area. On second thought, maybe either project could win a science fair contest.

Predicting Autumn Leaf Colors
Remains Unpredictable

The changing color of forests in autumn is a temperate zone phenomenon more predictable than hurricanes or election outcomes. Grammar school children know of its coming as evidenced by bulletin boards of colorful, leaf-shaped paper cutouts. Each year, thousands of people visit the Great Smoky Mountains National Park to admire the spectacle of autumn trees. Even a walk through local woods can offer a many-hued picture for anyone to appreciate.

But what determines the color of tree leaves in the fall, and why is predicting exactly when fall colors will appear at any particular location so difficult? Why do colors vary so much from year to year and place to place? Botanists are a long way from predicting accurately the timing and regional occurrence of fall colors in a given year or how vibrant the colors will actually be. Despite what you may read about the best time to view fall colors, no one knows for sure.

We do know that temperate zone forests change color in autumn. And thousands of leaf peepers make an annual pilgrimage to view trees in the Smoky Mountains and elsewhere, despite not knowing exactly when or where fall colors will be the most spectacular or whether colorful leaves will even still be on the trees.

Why do some trees have brown leaves, others have brilliant red or yellow ones, while still others stay green all year? Three basic pigments are responsible for most annual color patterns in plants. Chlorophyll, which makes leaves green, is the dominant pigment. As the days get shorter, some leaves display two additional

Q: I live in Atlanta and see geese flying very low over our neighborhood, sometimes early in the morning, sometimes in the late afternoon. Twice during autumn I have seen flocks of geese flying north. Do you have any idea why the geese would be headed in the "wrong" direction? Shouldn't they be migrating south at this time of year?

A: Most Canada geese seen in Georgia and elsewhere in the Southeast are year-round residents that do not migrate at all. They prefer to spend the night on the still water of a reservoir, lake, or pond, but because they eat grass, they forage during the day on grassy expanses, including golf courses and parks. Most of these resident, nonmigratory geese simply fly back and forth between foraging locations and roosting sites. Therefore, they could be flying in any direction, and the distances they travel daily could be just a few miles or up to a hundred miles or more.

pigments: carotenoid, which results mainly in yellow and orange, and anthocyanin, which results in red. Carotenoids absorb energy and help protect leaves from sun damage. Anthocyanins are less well understood but may also reduce sun damage and deter fungal pathogens. As far as leaves falling, botanist Linda Lee of the Savannah River Ecology Laboratory notes that "autumn weather per se is not what kills the leaves; the tree itself does, by pulling nutrients out of leaves and sealing them off. Once the leaf is an empty shell, it falls. In other words, the tree sort of beats winter to the punch."

The pigments that produce the variety of fall colors in tree leaves are themselves at the mercy of three primary environmental factors—temperature, day length, and rainfall. All are critical in determining how pigments express themselves. Cool autumn temperatures cause chlorophyll to degrade in many deciduous trees. Thus the carotenoid and anthocyanin, normally masked by the chlorophyll, are accentuated. Colorful displays of reds, yellows, and oranges are created. But a sudden, heavy frost may break down the accessory pigments as well as the chlorophyll, which means colors are muted.

If autumn cooling is gradual, colors may be dull because chlorophyll remains in the leaves, preventing full expression of the brightest yellows, reds, and oranges. To further complicate predictions, pigments respond differently to temperatures based on the timing and amount of rain during previous days or weeks. Predicting exactly when fall colors will appear in a particular year and what they will look like is as complicated as forecasting the weather—and just as unreliable.

Heavy summer rainfall can have a profound effect on fall foliage as the plants will produce more sugars and more pigments. If chlorophyll is broken down rapidly, the remaining colorful pigments will be at highest density, creating intense autumn colors and delighting the leaf peepers. A wet summer means more leaves, more leaf surface, and therefore more color displayed in the forest.

The geographic region and the types of trees also have a major influence on leaf color. Brown autumn leaves characteristic of many trees in warm regions of the South are often a result of pigments associated with tannins, which mask the red and yellow colors. Tannins are believed to discourage insects from feeding on the leaves. Evergreens, such as pines and magnolias, are another special case. Many evergreens have tough, waxy leaves and less watery sap that allow them to withstand extreme winter cold. The chlorophyll remains visible in such trees and other pigments are seldom expressed.

The biological purpose, if any, of the colors produced

by leaf pigments remains unresolved. Many questions about leaf color have yet to be answered, and scientists may never be able to predict exactly when and where fall colors will be at their most intense. Nonetheless, each year people will venture out to admire Mother Nature's fall spectacular, a pastime that will always bring its own rewards. The more we appreciate the complexities of nature, from all perspectives, the more likely we are to want to preserve and maintain the delicate networks that make up our natural world.

Everybody Recognizes a Hornet's Nest

A few things in the environment are so distinctive that no one is likely to confuse them with any other natural object. I saw three of them during an autumn walk in the woods: a box turtle, big red mushrooms, and a hornet's nest. Fortunately for me, my grandson Parker noticed the hornet's nest and pointed it out before I collided with it. The two-foot-long, football-shaped paper structure looked like a piñata hanging head high in the woods.

The first step to enjoying a bald-faced hornet's nest is to be aware of it before you bump your head on it. These members of the yellow jacket family will protect their nest by repeatedly attacking and stinging anything perceived as a threat. Our German shepherd once ventured too close to a nest and half a dozen winged black-and-white defenders swarmed out of the nest opening like bullets, all finding their mark. Last week my grandson

Q: Two weeks ago, our goldfish pond had hundreds of little black tadpoles swimming around in it. Now only a few are left. Did the goldfish eat them?

A: Goldfish are in the carp family, and most eat detritus, algae, and other plant material. They probably do not eat tadpoles in large quantities, although they might eat frog eggs and an occasional tadpole. Some toads have tadpoles that can develop from eggs to toadlets over a two-week period, so they probably just metamorphosed and left your pond. Or an enterprising watersnake could have found its way to your pond and enjoyed dining on tadpoles.

and I watched from about fifteen feet away as hornets landed at the entrance and entered the nest while others were coming out. I thought about pitching a little stick to jiggle the nest and see what would happen, but I was overcome by a wave of sanity when I realized I wasn't positive I could outrun my grandson.

Bald-faced hornets are the largest North American yellow jackets. An entomologist would be quick to tell you that they are not true hornets like the European and Asian species, but let's call them hornets anyway. They are certainly large enough, being almost an inch long, but they do not sport the characteristic black-and-yellow banding of the smaller varieties. In flight they look mostly black with light markings. Head-on the face looks like a fierce mask of ebony and ivory. And they don't just look ferocious. Bald-faced hornets feed not only on nectar, pollen, and tree sap but also on insects, including large ones such as cicadas and praying mantises. The most ambitious predatory takedown for one of these hornets, documented in British Columbia, was a rufous hummingbird! The geographic range of the species includes all of the contiguous United States and southern Canada.

The life cycle of this fascinating animal is complex in some ways but relatively straightforward in others. In early spring, when sustained warm weather appears certain, female bald-faced hornets emerge from winter dormancy and each selects a nest site. Some are low to the ground like the one we found recently; others may be more than fifty feet high in a tree. The female builds a small wasplike nest and lays an egg in each cell. The hatching hornets are all infertile females that immediately begin expanding the nest by chewing wood and mixing it with their saliva, which acts like starch on a cotton shirt. Using their legs, they begin shaping what will become the hive.

The queen meanwhile lays more eggs, producing a larger workforce. The process continues until the nest is finished. The interior of the large oval hornet's nest has several layers of what look like standard-size nests that paper wasps build under the eaves of houses. They will serve as home for the developing larvae and the colony of around four hundred. As autumn approaches, the queen alters the egg-laying process to produce fertile offspring—males, called drones, and females that will be the future queens. Mating occurs before cold weather sets in, producing new queens that seek hiding places beneath ground and in rotten logs and tree cavities. At year's end the queens are the only survivors; the workers and drones have died.

A hornet's nest is an amazing piece of natural architecture that can be collected and preserved without harming nature. By the first frosts, all the workers have perished, the queens have departed, and the unattended nests will soon be damaged by winter winds and rains. The queen hornets do not come back to reuse an old nest, so once the nest has been abandoned, it is quite acceptable to remove it from its environment. Be sure to pick a cold day, because any remaining female workers will defend the nest till the very end. Drones are not a problem; they have no stingers. When cold weather arrived, the hornet's nest we found made a great show-and-tell at my grandson's school.

Snakes Are Much in Evidence in Autumn

Most people appreciate autumn's cooler temperatures and fall colors. I personally like fall because more snakes are more abundant than at any other time of the year. North American snakes actually occur in greater numbers in the fall than any other time of the year and often are more visible because vegetation cover is reduced. So . . . because I like snakes, autumn is here, and people are more likely to see snakes, I feel justified in writing once again about these often reviled but quite important components of our natural environments.

Most US snakes are born in August or September. A few rush the season and are born or hatch in July. But due to natural deaths and a few that are intentionally caused by humans, the actual numbers of every snake species decrease each month from autumn of one year to the next. Ratsnakes, corn snakes, kingsnakes, and racers are among the kinds that lay eggs in early summer; these hatch in late summer. Rattlesnakes, copperheads, cottonmouths, and watersnakes hold their babies in the body and give live birth during the same season.

Baby snakes often make their debut around houses as they search for their first meal before cold weather arrives. Another reason for the prevalence of snake sightings in the fall is that both adults and young begin searching for safe hiding places to spend the winter dormancy period. Because they are more active aboveground than normal, they are more likely to be seen. Due to natural mortality rates that affect any animal species, the actual population size of every snake species begins to decrease by midautumn. By spring, most of them have been consumed by predators or died in other ways.

Some species, such as canebrake and timber rattlesnakes, mate in the fall and can often be seen crossing roads. The males, which get larger than females, are the ones that are most likely to be seen as they are the ones on the wide-ranging quest for receptive females. Research studies have documented that male canebrake rattlesnakes, as well as most other snakes for which such information has been taken, are killed in much higher proportion on highways than are females.

I receive many reports of autumn snake sightings. A typical snake inquiry is generally from someone who has just had a close encounter. People usually just want to be assured that the snake in their yard is not one of the venomous species. I am wary of identifying a snake based on someone's verbal description, even though I may think I know what it is. For example, a snake in an attic is nearly always a ratsnake and would almost never be

Q: If pythons were to enter other southern states, would they become a major problem?

A: The presence of occasional individual pythons in areas outside of Florida already occurs but is unlikely to become a major environmental or safety problem. The media will thrive on the sensationalism of reporting that a giant snake is around, but as far as being a danger to people and pets, the pythons would not be any more of a threat than lots of other potential but rare wildlife hazards (coyotes, venomous snakes, mountain lions). In colder climates outside Florida, pythons could be inactive for up to half the year.

a rattlesnake, cottonmouth, or copperhead. But I do not want to declare that it is unquestionably a ratsnake. Exceptions in biology are all too common, and I would not want to have someone bitten by the one copperhead in a thousand that decided an attic was a nice place to hide. Digital photography has been a significant advancement in snake identification. An email with a brief description of location and habitat accompanied by a photo of the snake itself is usually all that is required.

I do not claim that all snakes are harmless. Clearly, some protect themselves with fangs and venom, and under certain circumstances people end up as the victims. So, yes, some snakes can hurt you. But so can some dogs. And I can say on behalf of snakes that, just as with dogs, you usually have no cause for alarm. Snakes are not out to hurt or bother anyone—they just want to be left alone to find food, another snake, or a hiding place. Rest easy knowing that no venomous snake in North America will intentionally pursue a person, although some will stand their ground and even strike if you get too close. But they never ever come looking for you.

Our natural environments, which include snakes, are priceless. Because of people's fascination with and fear of these sinuous reptiles, snakes serve as a barometer of the public's mindset toward wildlife and natural habitats. Attitudes about snakes are one measure of the extent and effectiveness of environmental education in a region. The simplest rule for anyone who does not like snakes is to leave them alone. But a more productive approach is to get to know a herpetologist or at least someone knowledgeable about and comfortable with snakes. The more you learn about our native wildlife and their habitats, the more you'll enjoy those autumn walks, even if it's just a stroll around your own backyard.

One of our more memorable autumn encounters in late August started with an outstanding encounter in the woods, and by mid-September it got even better. The first occurred when my grandson Nick, son-in-law Keith, and I found an enormous canebrake rattlesnake stretched out in front of a tree stump in the woods. It rattled at us, halfheartedly, and then languidly crawled into a large hole beneath the stump. It was in the same place on three more visits over the next two weeks. Cool enough, but the fifth visit made it really special—stretched or coiled around the stump were five baby canebrakes with the big one.

Rattlesnakes are livebearers and do not lay eggs, so the mother is around, intentionally or not, to protect her young when they are first born. Within a day or so after birth newborn snakes generally shed their skin. Sure enough, when we went back three days after seeing the babies we found five shed skins around the stump—and no snakes. The thrill now is that we know that at least six canebrake rattlesnakes are roaming around in the woods we enjoy walking through.

Admittedly, some people will probably never learn to acknowledge snakes as an acceptable component of our natural habitats. But such narrow-mindedness is dwindling as society becomes more educated about all our native wildlife species and more accepting of the minor risks and major benefits that accrue to protecting them and their habitats.

Turkeys Are Here to Stay

Turkeys always start their celebrating the day after Thanksgiving. They are totally unappreciative that each year our country dedicates a day to them. This Thanksgiving Americans will consume more than 45 million turkeys. Those ending up on dinner tables amid dressing, gravy, and cranberry sauce will mostly be the fat, white-feathered, commercially raised birds that folks will pick up from local grocery stores. A few, however, might be the original native species, the wild turkey, acquired through legitimate hunting. The nonmigratory wild turkey is familiar to all Americans. Even schoolchildren can draw a picture of a turkey.

Today, wild turkeys thrive throughout much of their original geographic range. A lot of the credit for the reestablishment of the species over most of the country can be attributed to efforts by the National Wild Turkey Federation (NWTF) headquartered in Edgefield, South Carolina. According to their website (www.nwtf.org), NWTF, founded in 1973, is "a nonprofit conservation organization that works daily to further its mission of conserving the wild turkey and preserving our hunting heritage."

Turkeys occurred naturally in the 1700s from southern Canada into Mexico and Guatemala. Based on reports of early naturalists, eastern North America had an abundance of wild turkeys. As the country developed, the species began to decline and gradually disappear. Old reports state that the last wild turkey in Massachusetts was killed as long ago as 1851. Some probably survived in forested areas in most states south of New England, although populations became sparse everywhere. Into the 1950s many veteran ornithologists had never seen a turkey in the wild because of their rarity.

If wild turkey population levels had continued on the trajectory they took following the arrival of the first European settlers, the species might have been extinct by the end of the twentieth century. We would likely be remembering turkeys the way we do passenger pigeons and Carolina parakeets—as an extinct species whose disappearance could easily have been prevented. Fortunately, wild turkeys have returned to the landscape, and their numbers today far exceed those of only a few decades ago.

The present-day success of the wild turkey clearly demonstrates how regulated hunting can have a positive effect on a popular game species. Despite complaints by antihunting groups, a species favored for sport hunting often fares better than other native species. One reason is that substantial efforts are made to protect game species, including maintaining suitable habitat conditions and tightly controlling illegal hunting. Federal and state support for game management comes directly from taxes paid by hunters for firearms, ammunition, and other hunting essentials. An additional feature of focused attention on a game species is that scientific research is conducted to understand its behavior and ecology in all seasons and under different environmental conditions. Environmental restoration, research, and management programs to ensure suitable wildlife habitat and population viability for turkeys have been coordinated by NWTF for almost forty years.

As NWTF notes in brochures and on their website, wild turkeys are one of the country's great conservation success stories. A reliable estimate is that around

10 million wild turkeys were in the United States "before the settlers arrived" and that the big birds served as "an abundant and important food source." According to some estimates, by the 1930s only 30,000 to 100,000 wild turkeys remained across North America due to unregulated hunting and habitat destruction. That would mean 3 to 10 birds for every 1,000 that were once present—a dramatic decline. More than 7 million wild turkeys are now found "throughout North America, thanks to the efforts of state, federal, and provincial wildlife agencies, and the support of the NWTF, its members and partners." That partnership conservation effort is admirable and highly successful.

One of the questions people ask about turkeys, most often around Thanksgiving is, can turkeys fly or do they always stay on the ground? Domesticated turkeys are generally too heavy to fly and probably can't run very fast either. However, wild turkeys are fast runners on land and will often try to escape a threat by sprinting through the woods. But wild turkeys can fly for short

Q: We live in Tuscaloosa, Alabama, in a section of town that was badly damaged by the April 27, 2011, tornado. Our neighborhood used to be quite shady because of all the huge old trees. Now we have no tree canopy at all. Some of our neighbors think we have an infestation of black widow spiders. Does that seem likely?

A: Black widows, especially the northern or woodland black widow (*Latrodectus variolus*), do prefer sunny light gaps in deciduous woods in the southeastern United States. The tornado created lots of light gaps that previously did not exist. There could well be a spike in the black widow population due to this phenomenon.

distances and most roost up in trees. A startling experience I had with turkeys was during a nighttime canoe trip on a creek on a moonlit night when we were not using flashlights. We bumped into a floating log, waking up a flock of wild turkeys roosting in the oak trees above. A dozen or more flew directly overhead with a whirring of wings and crashing of tree branches. When I finally caught my breath, I heard them in the trees on the other side of the creek resettling for the night.

Wild turkeys thrive throughout much of their original geographic range, with every state except Alaska having populations that are sizable enough for regulated hunting seasons. According to NWTF, the greatest numbers of wild turkeys are in the eastern half of the country. Among the highest are Texas with an estimated 500,000 and Alabama, considered a top choice for turkey hunting by the pros, with 400,000. Of course, 240 million turkeys will be raised commercially this year, but the idea of potentially encountering a wild turkey, part of our natural heritage, in almost any forest in the country is an exciting prospect.

As Thanksgiving approaches next year, let's spare a thought for the wild turkey, North America's largest game bird. Due in large measure to NWTF, more wild turkeys now roam the woods and fields of America than have done so since colonial times. True, more turkeys will end up on dinner tables than ever before as well (though most of those will be the domesticated variety). But whatever your family serves for Thanksgiving, we can all give thanks that wild turkeys have been spared the fate of the passenger pigeon and the Carolina parakeet.

Cranberries Are an
All-American Treat

When Americans celebrate the holiday season from Thanksgiving through Christmas iconic foods immediately come to mind—apple pie, pumpkin pie, sweet potatoes. And, of course, cranberries. But if you want a truly American meal, you can dump the first three, because only cranberries are native to the United States.

Apples? Whether the first ones came to our shores with the Pilgrims or with earlier emigrants from England, France, or Spain is uncertain. But that apples are native to Asia is an accepted fact. Was Johnny Appleseed distributing an alien invasive species? Apples taste good and remain edible for a long time, so the fruit rapidly became popular in colonial times. Also, apple trees grow well in many regions, and by the time of the Revolutionary War thousands of varieties had been produced by early agriculturists. Apples are here to stay but were not a major part of the holiday meals of the first settlers.

What about pumpkins? Pumpkin pie is considered by many Americans to be traditional holiday fare. According to botanical scientific literature, pumpkins were one of the many forms of squash derived from wild gourds that probably originated primarily in Mexico. They were already being cultivated when Columbus arrived in 1492, and 1542 records indicate that pumpkins were being grown in Europe.

Sweet potatoes, another standard for US holidays, may have the most questionable origins. Tropical America and Indonesia have both been suggested as the original home of the sweet potato. But in either case, like apples and pumpkins, they did not originate in what is now the United States.

Cranberries, which are related to blueberries, are a completely different story, being part of the native flora of eastern Canada, and the northeastern and north central United States. The woody vines grow naturally in acidic bogs. The original name "craneberry," eventually shortened to "cranberry," stems from the flower's resemblance to the head of a common and distinctive native bird, the sandhill crane. Early colonists would have been aware of the cranes, and the little flowers having what look like long beaks and red heads on a curved neck would be obvious. Check out pictures of cranberry flowers and sandhill cranes to see the remarkable similarity.

A plant that thrives in wetland habitats in cold climates is unlikely to be grown effectively in hot, arid regions. But in checking scientific journals to determine what had been discovered about cranberries within the past year, I was surprised to find many of the recent studies were conducted by scientists in the Middle East. These included ones from Egypt, Turkey, and Saudi Arabia in publications such as the *International Food Research Journal* and *Annals of Applied Biology*. One examined the beneficial health properties of fresh cranberry juice due to its antioxidant effects. Another was determining meteorological effects on flowering, and a third concluded that cranberry fruit extract improved the quality of white soft cheese. *Arab News*, a widely distributed newspaper published in Saudi Arabia, recently had an article extolling the "proven health benefits" of American cranberries. The popularity of cranberries has extended to the Middle East, despite the lack of suitable natural habitat.

Although the introduction of American cranberries to the Old World is gaining appreciation, not all is perceived as positive. One journal article was titled "A

New Alien Plant Species in Lithuania," with cautionary comments about guarding against "probable invasions in natural bog habitats." Apparently, eastern Europe's peat bogs and cold climate are comparable to that of the cranberry's native habitat. Intentionally or accidentally introduced cranberries can propagate and be dispersed by birds. Ironically, America's most popular native berry is considered a nuisance in some parts of the world.

Our only truly American fruits consumed throughout the country in large quantity during the holiday season are cranberries. The expression "as American as cranberry pie" is unlikely to replace the one referencing apples, but it would be more accurate.

Winter

Is Gift Giving Unique to Humans?

Do humans do things that no other animal does? To be sure, a few special types of behavior come to mind. Praying, exploring outer space, and being a lawyer are, as far as I know, three uniquely human activities. But we do few things that some other species does not also do. Even gift giving is not a practice unique to humans.

December 25 and 26 bring us two gift-giving days in a row—Christmas, then Boxing Day. The origin of the latter has almost as many explanations as mistletoe has berries, but there is general agreement that Boxing Day began in Britain and was marked by giving money or goods to the less fortunate. Many animals are gift givers. Think of your cat who has just presented you with a dead shrew from your front yard. What could be a more delightful offering? The lioness Elsa in *Born Free* brought her cubs to her former human protectors as a gesture of appreciation for their erstwhile care.

The practical aspect of parent birds feeding their young is obvious, but they are bringing gifts. Bower birds of New Guinea take gift giving considerably further. To attract a female for mating, the male uses twigs and vines to build a bower, an elaborate structure on the ground. Brightly colored berries and flowers are arranged around the bower to make it attractive to the prospective mate. The objects used to adorn the bower are not major food sources or otherwise useful to the builder or the object of his ardor. They are merely decorative, much like ornaments and lights adorn a Christmas tree.

Ants, termites, and other social insects demonstrate the spirit of giving to special members of the colony. The

Q: My husband and I were visiting friends in Clarksville, Tennessee, and saw an unusual sight: a squirrel and a rabbit playing together in the yard. I don't mean they were tossing a ball back and forth or taking turns pushing each other in a swing, but they were clearly engaged in playful behavior. The rabbit would hop and the squirrel would chase it; then the squirrel would run and the rabbit would hop after it. All four of us witnessed this one afternoon. The next day we observed the same behavior; presumably these were the same two animals we had seen before. These were definitely wild animals, not someone's pets. Is "playing" a common behavior in wild animals?

A: Anyone who has ever had a puppy or a kitten knows that playful behavior is common among domestic dogs and cats, and it occurs frequently among some wild mammals, generally being restricted to juveniles. Behavioral ecologists assume that play in juvenile mammals confers some biological advantage to the individual as an adult, but for most species no research has been done to confirm this idea. However, in a study of ground squirrels, scientists documented that playing by juveniles enhanced the motor skills of individuals as adults that led to better avoidance of predators, more success in fights, and, for the females, more surviving offspring.

queen of many species sits in her parlor day and night producing young while accepting a steady supply of edible gifts brought to her by the workers. Extreme gifting is exemplified in the spider world by a male that offers itself as a meal for the female with which it has just mated. Clearly, some gift giving can go a step too far.

We once observed an unusual and impressive act of gift giving by our dog Gilbey. The story of his extraordinary behavior seems to me to bear repeating during this holiday season. Gilbey was not noted for aggressive behavior. Though he was large and looked threatening, like a long-toothed Doberman-Rottweiler mix that might rip your throat out if you smiled crooked, in reality Gilbey was more like a kitten in a dog suit. His role as watchdog was to bark so that Nero Wolfe, our real dog, knew that danger lurked. I'm not sure what Gilbey would have done if actually put to the test.

But even for a dog in kitten's clothing, Gilbey demonstrated remarkable behavior when friends brought their two-month-old baby for a visit. They put the baby on a blanket on the floor to do whatever babies do while grown-ups are talking. When the dogs came into the room, each gave a disinterested sniff at the baby and then went about doing whatever dogs do while people are talking. But in a few minutes Gilbey returned with a dog bone in his mouth (not the bone of another dog but the kind that comes in a box from the grocery store). Knowing how ardently he guarded these valuable objects from the other dog and the cat, we were amazed to see him walk over to the baby and drop the bone on the blanket. He then moved back a foot or so, lay down, and watched. Without any doubt, he had presented the baby with a gift. I think this kindly act was a holdover from the dog's ancestral past wherein wolves, the highly social ancestors of dogs, qualify as gift givers by sharing their prey with other members of the pack. Gilbey clearly still had some wolf in him.

Who knows what the story is with domestic pets, but one common feature of gift giving among wild animals is that the behavior is not altruistic. Any animal giving a gift is doing so for a self-serving reason—to raise its young, to attract a mate, to survive by a division of labor among a group's members. The gifts have a practical value for the giver. A male common tern brings the largest fish he can catch as an offering to the female he is courting, presumably to show off his fitness as a mate.

As humans, we have many animal traits that have been molded by culture over the ages. Sometimes the true origin and ancestral purpose of a trait has become lost in the evolutionary past so that we think of ourselves as having risen above the rest of the animals. Maybe we have in some instances, and perhaps gift giving is one of those situations. Let's hope that for most of us giving gifts is prompted by the altruistic belief that to give truly is better than to receive.

Deck Your Halls with Boughs of Holly

Answers to these questions I have received about holly are perfect for the holiday season.

Q. Are holly trees native to North America? How big do holly trees get? Where did the idea of using holly at Christmastime originate? Why do holly leaves have those needle-like spines on them?

A. The American holly, scientific name *Ilex opaca*, is native to the eastern sector of the United States from New England to the upper half of Florida to eastern Texas. Hollies also occur naturally in Pennsylvania, West Virginia, and southern Missouri.

Nearly everyone knows what a holly tree looks like, but they can reach sizes that may surprise you. American Forests, an outstanding conservation organization, has a National Big Tree Program that identifies the largest individual trees of every species in the United States. Several champion-size holly trees vie for the record. A tree at the Chelsea Historic Site, Maryland, is a cochampion because of a combination of criteria (height, trunk circumference, and crown spread). The tallest holly, in Alexandria, Virginia, measures 68 feet from the ground to the top branch (about as tall as a six-story building). American Forests's website lists a holly in Chattooga County, Georgia, as the one with the largest circumference (148 inches), whereas the fall 2012 edition of the *National Register of Big Trees* cites a tree in Arlington, Virginia, with a girth of 154 inches. In either case, two tall men could barely reach around the trunk and touch their fingers. American holly trees grow slowly, but they can get really big.

A closely related species, the European holly with the scientific name *Ilex aquifolium,* became associated with the Christmas season centuries ago. Like the American holly, the European version has glossy green, waxy leaves and bright red berries in winter. A hardy plant whose branches could be brought indoors to liven up a bleak winter day, holly was probably first used as decoration by pagan cultures, including the Druids. Christians also liked the holly's cheerful display and before long it became a common decoration in many parts of the world, including North America. Today, cultivated varieties of the two species are commercially produced in many regions.

From an ecological perspective, European and American holly trees have a common trait—a holly tree is either male or female, and the two sexes differ dramatically. Both have flowers, but only the female trees produce the bright berries; the male trees simply have the shiny green pointed leaves. Bees, wasps, and other insects pollinate holly trees, and unless a male tree is nearby, the fruits, that is the berries, of the female tree cannot develop. So, even though the female tree is the one we consider most impressive and we prefer for our Christmas wreathes, no berries will appear unless an unassuming male is in the vicinity.

As for the ecological explanation for why holly leaves have those needle-sharp points on the end and sometimes on the sides, the scientific jury is still out. Some grazing animals might be deterred by having to bite around a pointed spine, so one idea is that they prevent animals such as deer from eating them. One proposal, which I checked out on a holly tree in my backyard, is that the lower leaves have more spines than the upper ones—the supposition being that grazing deer would be more likely to eat the lower leaves. Indeed, on my holly tree, most leaves ten feet above the ground have fewer spines, but we need more data than my casual examination of a single tree. Another idea is that the spines are a perfect conduit for freezing rain to run off during an ice storm, a useful trait for an evergreen tree like holly that might be damaged if weighted down by ice.

Whatever the explanation may be for the pointy leaves, a few branches of holly, with or without the colorful berries, can add a festive touch to your holiday decorations.

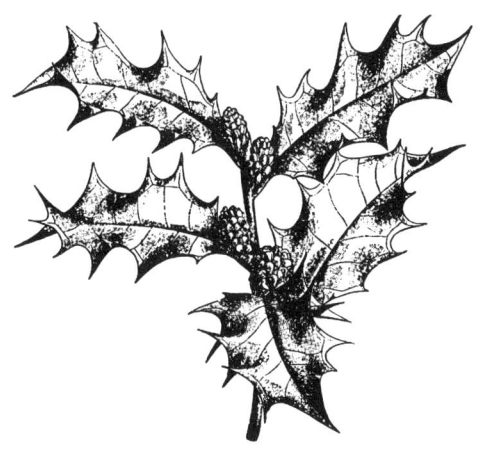

Mistletoe Is America's Most Popular Parasite

Mistletoe is the only plant associated with Christmas that has flowers pollinated by insects, has seeds transported by birds, takes its water and minerals from trees, and is not displayed in church. "Mistletoe" refers to any of more than two hundred species of semiparasitic shrubs found worldwide. Mistletoe lives throughout the southern United States, from the Atlantic Coast to California, and on every continent except Antarctica. Like true parasitic plants, mistletoe is devoid of roots. Instead, the dark green shrub has extensions called holdfasts that grip the host tree. These rootlike anchors suck water and nutrients from the tree. Thus, mistletoe is only found on living trees, which are essential to the mistletoe's survival. In contrast, Spanish moss uses a tree, dead or alive, only for support, extracting water and nutrients from the atmosphere.

In the South, tiny yellow flowers bloom on the evergreen mistletoe from fall to winter. The familiar white berries, which begin to form soon after pollination, resemble little packets of glue around tiny indigestible seeds. A mistletoe plant can be either male or female and, like a holly tree, only the female plant has berries. Eating mistletoe berries may make a person ill, and some say they can be potentially lethal for humans, but birds seem to be unaffected by any toxicity.

Birds' immunity to mistletoe's poisonous qualities is essential to the welfare of the plant. The dispersal and propagation of mistletoe is largely dependent on birds that eat the berries but do not digest the seeds. Ecological studies suggest that seeds are most likely to survive

Q: I live in northern Alabama and recently saw a dozen lizards (the chameleon-like ones that change from brown to green and back) in the yard. Is there some kind of food we could put out for them during the winter or do these lizards hibernate?

A: Green anoles hibernate in colder regions of their geographic range, which extends from the Carolinas, throughout Georgia and Florida, to Alabama and on to east Texas. They often hibernate in large groups and were probably congregating before it turned cold. Anoles spend winter under bark, inside rotten logs, and under boards of houses and barns. They can be seen on bright, sunny days in winter basking in the sun. As for feeding them, they will do fine with no help from us as they eat little or nothing in winter. When spring arrives, they will feed on small insects and spiders around the yard.

and grow if a bird deposits them on the same species of tree on which the parent plant lived. During spring migration, a flock of berry-eating cedar waxwings can result in newly developing mistletoe plants being far away from where the seeds were ingested. Mistletoe thrives in bright sunlight in the uppermost branches of big oaks and is typically absent from pines and from evergreen hardwoods such as magnolias with needles and leaves that would shade the mistletoe.

A parasitic lifestyle is unusual among flowering plants. Nonetheless, many aspects of mistletoe ecology are well understood. Competition to obtain water, minerals, and even space itself is highly intense among most plants, but mistletoe does not encounter such problems. Tree limbs are a ready source of water and minerals for this unusual little plant, and its absence from the uppermost branches of a tall oak is probably because no bird has dropped a seed there, not because of competition with other mistletoe plants.

Over the ages, mistletoe has been credited with some intriguing qualities, perhaps, in part, because of its many unusual ecological properties. According to Scandinavian legend, mistletoe was the only organism in the world from which Odin's son Baldur was not protected, and a mistletoe dart was the cause of his death. Mistletoe is also associated with the Druids, the mysterious, oak-worshiping sect that inhabited the British Isles centuries ago. The Druids considered mistletoe a plant of honor and power. According to legend, when the plant was found growing in an oak tree, the Druids performed sacrificial ceremonies at the tree on the sixth day after a full moon. The Druids reportedly used a golden sickle when harvesting mistletoe from a sacred oak. Although the berries appear just in time for Christmas, mistletoe is not used as decoration in churches, one reason presumably being because of its close association with the Druids.

Using mistletoe as a romantic lure was common in England at least as early as the 1500s. In 1520, William Irving wrote that a young man should pluck a berry each time he kissed a girl beneath the mistletoe. A version of that tradition persists today in secular decorations around Christmas. And though mistletoe may be excluded from wreaths and floral displays in churches, it will be found in many a home during this season, hanging in a doorway and enticing people to exchange a holiday kiss.

Where Would You Find
Twenty-Two Turtle Doves?

The most ecologically relevant song of the Christmas season is the one that starts off with a partridge in a pear tree as the first of many gifts. According to my calculations, by the twelfth day of Christmas someone's true love had delivered more geese and swans than any other bird. In fact, 42 of each ended up under new ownership, compared to only 12 partridges and 22 turtle doves. At 30 and 36, respectively, French hens and calling birds were closer in numbers to the geese and swans but still fell short. I am almost certainly not the only person to have made these Christmas bird counts, so if I got them wrong, please let me know.

In case you are wondering what this has to do with ecology and the environment, remember that anything involving an animal, plant, heavy metal, air, water, sound, most other tangible things, and many intangible ones can be construed as having something to do with the environment. Since this is the holiday season—a time for giving—my gift to readers is to apprise them of some ecological aspects of this song. For example, how many kinds of geese and swans are there? Are all of them suitable for gift giving? What are the benefits of receiving a partridge or a French hen? What in the world is a swan goose, and is it suitable for gift giving? What's another name for an ugly duckling? And does anyone really know what a calling bird is?

The so-called swan goose could create a dilemma for someone giving geese and swans to a true love. It is genetically related to both geese and swans but not enough so to be classified as one or the other. Do you give them as geese on the sixth day or swans on the seventh? This isn't likely to affect US gift givers as the species is rare throughout most of its native range from Mongolia to Korea.

All swans and geese belong to the same family of mostly migratory waterfowl. For your gift on the sixth day of Christmas, you can select from more than a dozen kinds of geese found worldwide. Several are native to the United States. The best-known ones are the snow, blue, and Canada geese, which travel up and down much of the country during migratory seasons. One goose that would not be suitable for gifting is the Hawaiian goose, or nene (pronounced "nay-nay.") The Hawaiian state bird is on the federal endangered species list. Only thirty nene are believed to have been alive a half century ago, although successful recovery plans have increased the numbers. Nene do not migrate. (Why fly to another island with the same temperature?) They mostly walk around on lava slopes instead of swimming around in water and have less webbing on their feet than other geese.

To most Americans a swan is a big white bird that looks regal gliding around on a lake, and indeed most of the world's half dozen or so swan types are all white or mostly so. The species native to Australia, however, is black with a bright red bill. I doubt that 7 black swans have ever been given to anyone on the seventh day of Christmas. As with many domesticated animals, male, female, and baby swans have different names. A male is a cob; a female, a pen. The babies, a.k.a. ugly ducklings, are called cygnets.

Some sources suggest that the 5 gold rings refer to rings on a pheasant's neck, which would make the first 7 gifts avian related. But the idea (however logical it might be) isn't likely to catch on at this late date, so I'm not including ring-necked pheasants in this Christmas bird count.

The partridge and the French hen are related to quail and pheasant. They can be pretty; they produce eggs that are edible; and they themselves are edible. Clearly, good gifts. Turtle doves, a type of migratory pigeon found from Europe to Africa, would be a poor gift choice because their numbers have declined considerably in recent years. "Calling bird" is apparently a corruption of "colly bird." Colly means "black or sooty," so the fourth day's gift would be blackbirds. Since according to Mother Goose they can be "baked in a pie," colly birds might be a nice gift.

In the song's final Christmas bird count, the record for highest number should probably go to the geese. All 42 of them were a-laying and would presumably soon have goslings running around, whereas the 42 swans were simply a-swimming, with no ugly ducklings expected. What someone did on the twelfth day of Christmas with 184 mostly big birds running around their house (or sitting in pear trees), I can only imagine. Maybe the 48 maids that I calculate were present, stopped a-milking and started a-mopping.

What Is the Story behind
Poinsettias?

In checking into facts about America's most popular
Christmas flower, I was reading an older edition of
the Encyclopedia Britannica about a man described as
a "statesman noted primarily for his diplomacy in Latin
America" and also for being the first US ambassador to
Mexico. Joel Roberts Poinsett, of Charleston, South Carolina, was born in the late 1700s and lived till 1851. Only
near the end of the article was it mentioned that Poinsett
brought back from Mexico a flower that is now the top-
selling potted plant in the United States—the Christmas
flower, or poinsettia.

Although no one seems to dispute that Joel Poinsett brought the bright red flowers and their seeds to
the United States, some controversy surrounds the exact manner in which the species persisted and became
of commercial importance. The current thinking among
some horticulturists is that Poinsett sent the seeds to a
nurseryman named Robert Carr, who was married to
Ann Bartram Carr, whose grandfather John Bartram
established Bartram's Botanic Garden in Philadelphia.
Not surprisingly, this twisted tale has a connection with
William Bartram the traveler, who was the son of John
Bartram and the uncle of Ann, but whether William
ever saw a poinsettia, wild or potted, I have no idea.
Anyway, according to one of the largest producers and
distributors of poinsettias today, the Paul Ecke Ranch
in Encinitas, California, Colonel Carr "introduced [the
poinsettia] into cultivation and commercial trade . . .
[from] Bartram's Garden on June 6, 1829" (www.ecke.
com/company/historyofthepoinsettia/). The occasion

Q: Where I work in upstate New York, we came across a mudpuppy [salamander] in our steam condenser. What should we do with the critter? Is it safe to return it to the freezing river? We figure he made his way into a warmer pipe under our building, but possibly he got sucked in from the nearby river. Any help you can give would be appreciated. I'd hate to see this critter meet an untimely demise considering what he must have been through to end up where we found him.

A: Mudpuppies in your region of the country live naturally beneath frozen lakes and rivers during the winter. So the animal will be fine if you release it into cold water. You should move a salamander, fish, or other aquatic animal from one aquatic situation to another in a manner that will have minimal thermal impact on the animal. In this case, put the mudpuppy in a plastic bag filled with the water you are currently keeping it in and place the bag in the river water for several minutes so the temperature changes gradually. Open the bag and release the mud-puppy when it is ready to swim out, preferably alongside a bank it can hide under to avoid predatory fish.

was an exhibition by the Pennsylvania Horticultural Society, which ultimately evolved into the long-standing Philadelphia Flower Show.

Poinsettias belong to a large and economically important family of plants known as the Euphorbiaceae, many of which are called spurges. The more than seven thousand species in the family are distributed throughout the continents and islands of the world, making it one of the largest and most widespread plant groups. Some euphorbia species are used as ornamental plants; others are the source of rubber, castor oil, and tapioca.

Poinsettias are similar to dogwoods in that the parts

of the plant that are so attractive to us are not actually petals; they are bracts, which are small, inconspicuous structures on most other flowering plants. Most of us are familiar with poinsettias as potted plants that are prevalent during the holiday season, but in their native Mexico and Central America, poinsettias are large shrubs that get more than ten feet tall. As with the commercial variety, these showy plants bloom during the winter in the wild.

Clearly, poinsettias have become a commercial product of note during the winter holiday season. We have all seen the traditional red poinsettias, which still constitute approximately three-fourths of the market. But varying shades of reds, pinks, creamy white, and a recently developed purple poinsettia are among the other varieties available today. The website for Cushman's Greenhouse has images for more than a dozen varieties in a range of colors, from Snowcap to Marble Jester to Freedom Salmon. Or you can Google "poinsettia colors" to see a host of photographs.

Many websites also offer tips on poinsettia care. Keep the thermostat set between sixty-eight and seventy degrees Fahrenheit, which is about normal room temperature for most people, being sure they never experience temperatures below fifty. Keep poinsettias away from fireplaces, heat vents, and areas where they might experience cold drafts. Water them whenever the dirt in the pot feels dry, but be careful not to overwater. The plant should be placed in an area where it can receive about six hours of sunlight each day, preferably avoiding direct sunlight. And if you want to keep your poinsettia thriving after the holiday season, use an all-purpose fertilizer. But, and this is very important, do not fertilize the plant until the blooming season is over. Properly cared for plants will last for several months.

Poinsettias are said to have a bitter, unpleasant taste and may cause an upset stomach but reportedly are not toxic (or at least fatal) to humans or (as was once believed) to dogs and cats. Of course, before relying on this presumed fact, the best policy with any plant that doesn't come from the food sections of the grocery store is look at it but don't eat it.

One explanation for the connection between poinsettias and the Christmas holidays is found in a Mexican legend. On Christmas Eve, so the story goes, a youngster (or a brother and sister in some versions) was prompted by an angel to pick a bouquet of weeds to take as an offering to the church. The child was ashamed to bring such a paltry gift but the angel insisted. In the manner of such stories, as the child laid the weeds before the nativity scene, they were transformed into brilliant red blooms that were the loveliest of all the offerings.

Today, J. R. Poinsett's fame rests primarily with his having introduced the United States to its premier Christmas season flower. But his name also lives on in some Spanish-speaking countries—in an entirely different context. Apparently his policies in Mexico were unpopular and his personality abrasive. Someone coined the word "poinsettismo," which is used to describe someone with "officious and intrusive behavior," a less flattering legacy than the beautiful "poinsettia."

Rudolph Is Not a Female Reindeer

In 1949, when singing cowboy Gene Autry said that Rudolph the red-nosed reindeer would "go down in history," he probably did not believe it. Especially as the song still remains the only one in musical history to be number one on the charts one week and completely off the charts the next. If Autry were alive today, he would also be surprised at the suggestion that Rudolph is a female. Despite continuing threats to small wetlands, challenges to the Endangered Species Act, and the debate about the causes of global warming, 'tis the season to be jolly. So with Santa Claus soon to be on his way we should at least set things straight on this contentious issue about the gender of Santa's reindeer.

The way Christmas stories abound in literature, television specials, and Hollywood movies, who would be surprised to find that some children these days think reindeer are as fictional as the Grinch, Frosty, or Rudolph? Well, reindeer are as real as caribou. In fact, according to some scientists, reindeer and caribou are the same species, although "Rudolph the red-nosed caribou" falls a bit short of the mark for lyrics. Those living in the Arctic tundra of North America are called caribou; those living in the same habitat from Europe to Siberia are called reindeer. Caribou thrive in the Arctic regions of Canada where the ground remains permanently frozen, but a few are found in northern Idaho, and large numbers live in Alaska.

Both caribou and reindeer belong to the deer family along with white-tailed deer, elk, and moose. All deer belong to a mammal group known as the even-toed hoofed animals, which include pigs, cattle, buffalo, and goats. More distant relatives are the odd-toed mammals, such as horses, zebras, and rhinoceroses. Only members of the deer family have antlers that are shed each year, rather than horns that persist throughout the animal's life. Reindeer and caribou are even further distinctive—females as well as males have antlers, leading to the idea that Rudolph may be a female.

The issue has arisen because female reindeer characteristically keep their antlers all year whereas males, like other deer, shed them and grow new ones prior to mating season. But before we bring about a metaphorical sex change in Rudolph, we will do well to remember that vast variability in traits, including antlers, can exist in the animal kingdom. Hence, some female reindeer do not have antlers, in spite of the general rule that females do. Also, the period of time over which antlers are dropped by deer varies greatly, with at least some individuals having antlers at almost any time of year. Wild reindeer and caribou mate in the fall, with males engaging in contests and using their antlers as weapons. So Rudolph can retain his traditional role as a male reindeer that, for whatever reason, keeps his antlers until December 26.

A nonbiological difference between caribou and reindeer is that reindeer are domesticated throughout much of their geographic range whereas caribou are wild animals hunted by people (another reason why Rudolph is not the red-nosed caribou). In fact, reindeer are the only members of the deer family to be successfully domesticated, providing meat, milk, and cheese, as cattle do elsewhere. Reindeer are also used to pull sleds. You'll have to ask Santa how they learn to fly.

Wild reindeer and caribou are noted for forming

Q: During a warm spell last week, my dad discovered some newly hatched painted turtles, which he proceeded to pick up and put in a bucket on our kitchen counter. I did some research and found that baby turtles, particularly painted turtles, will survive just fine during the winter in our large pond. The turtles have been living in a shallow dish with about one inch of water and a few rocks for three days. My question is this: Should we return the hatchlings to the pond to live out the winter, or would they have a better chance of survival inside in a dish or tank?

A: The hatchling turtles will be okay if you put them in the pond for the winter. A heavily vegetated area next to the shore would probably be best. They will find places to hibernate during the coldest periods and then emerge during the spring and start feeding. The threat of predators always exists for wild turtles but that's part of being a turtle.

large herds and making long-range annual migrations. Their herding tendencies and constant travel are a necessity during winter. They continually paw through Arctic snow to find food in the form of grasses and lichens known as reindeer moss. As the herd depletes the forage in an area, it must be on the move again.

A large herd offers protection against natural predators—wolves. An animal traveling alone would be easy prey for a wolf pack, but predators have difficulty

surprising an enormous herd moving over frozen terrain. Hundreds of eyes, ears, and noses (none of which are red) provide an early warning system, and wolves usually concentrate their attacks on stragglers that are sick or weak. Some other threats to the caribou of North America include grizzly bears and lynx that will attack calves in the summer. Caribou have their young in late spring, and a healthy baby is on its feet and able to run within an hour after birth.

A serious parasite common in white-tailed deer has spread into caribou populations in some areas. One speculation is that the clearing of the continental forests forced white-tailed deer farther north into the range of caribou, exposing them to the parasite. Another cause of high mortality in some years is simply starvation due to unusually long, cold winters with heavy snowfall. Sport hunting reduces the number of caribou in some areas. Fortunately, Rudolph and his eight reindeer followers (most with names that could be either male or female) do not have to deal with the likes of wolves, finding food in the snow, or the other perils of living in the wild.

Another song that Gene Autry popularized in the 1940s was "Here Comes Santa Claus." And both the jolly old elf and his red-nosed male reindeer will be here soon.

What Is a Groundhog?

So is cold weather nearly over for the year or can we expect six more weeks of winter? On February 2, known to most as Groundhog Day, Pennsylvania's Punxsutawney Phil will give his answer. Don't try to figure out why winter should continue for exactly six weeks or end more quickly, just enjoy the tradition. And certainly don't try to make any sense out of the notion that a giant rodent that lives in a hole in the ground and comes out to look for its shadow in February would be any better at predicting the weather than the weather app on your smartphone.

Though a discussion of winter weather prognostication, a centuries-old tradition, would be interesting, a look at groundhog ecology seems more suitable for an environmental column. The book *Mammals of Alabama* (University of Alabama Press, 2014) by Troy L. Best and Julian L. Dusi answers the ecological questions people might ask. Because Alabama has such a high diversity of native mammals, the book covers the biology of most mammal species in the country, especially for the eastern states. A map shows the geographic range of groundhogs: from Alaska, across southern Canada and the northern states, to New England and southward to most of the Southeast.

Groundhogs, also known as woodchucks or whistle pigs, are rodents in the squirrel family, which also includes chipmunks, tree squirrels, and flying squirrels. Groundhogs are in a group known as marmots, their closest relatives being yellow-bellied marmots (rock chucks) in western states. Like other rodents, groundhogs have two upper and lower incisors in the front of the mouth but no canines. The powerful front teeth can gnaw through any type of plant material, including bark, roots, and hard-shelled nuts. Their teeth grow throughout their lives. Without constant gnawing to keep them trimmed back, they would become longer and longer.

According to *Mammals of Alabama*, groundhogs "are vocal animals and may squeal, chatter, bark, or give a loud, shrill whistle." If an animal feels threatened by a predator in the vicinity, the whistle presumably serves as an alarm system for other members of its immediate family. A groundhog will typically head for one of the holes leading into a burrow, which can serve as protection not only from predators but also from wildfires and inclement weather. Burrows often have ten or more entrances on the surface.

Q: Please don't think me stupid, but an acquaintance told me that if I can't feed my ten-foot python, I should put it in the freezer for a month until I can afford some food for it. All I want is advice contradictory to this idiotic statement and to stop any cruelty from being inflicted.

A: Your acquaintance may be well meaning, but following that advice will leave you with a dead python. For many pet reptiles like pythons that are native to the tropics, cooling them down a few degrees below room temperature, but nowhere near freezing, may reduce their tendency to eat and lower their metabolism so that they use less energy. Thus your python might be able to go longer without eating and with no harm. However, although almost any snake can go a month or more without eating at low temperatures, a well-fed snake will be healthier. Also, be sure to keep water available even at lower temperatures.

Groundhogs are true hibernators. Prior to winter they store body fat and do not eat until they emerge in early spring. Based on their diet, they are primarily a combination of herbivore and insectivore. They eat nuts, leaves, and other vegetation, as well as a variety of insects. They also eat snails (gastropods).

Males will fight in the springtime during mating season. Although groundhogs are the largest members of the squirrel family in the eastern United States, they have no problem climbing a tree, whether to forage or to use a high perch as a lookout post. They can swim if necessary to escape a predator.

As for the groundhog's physical appearance, most people would agree that whistle pigs, especially the roly-poly babies, would easily qualify as cute. But a wild one would undoubtedly let you know with a solid bite that it did not want to be picked up. Even without canines, that would hurt. Besides the threat from their big front teeth, another reason not to pick up groundhogs is that they can contract rabies, although I do not know of any documented case of transmission to people. A squirrel on steroids might be entertaining to have around, but vegetable-eating groundhogs are not popular with home gardeners, and groundhog burrow openings can be a problem for a horse that steps into one.

If you are unconvinced that a hibernating rodent can signal whether it's time to put away your winter coat, rest assured you are not alone. Not everyone believes Punxsutawney Phil can really predict whether winter will continue for another six weeks. You might as well rely on the *Farmers' Almanac*.

MLK Day Offers Opportunity
for Environmental Lessons

I once gave a talk that addressed the question "what does Martin Luther King Day have to do with environmental attitudes." Dr. Martin Luther King Jr. was a great and special man, a teacher. And some of the lessons he taught work as well for ecology as they do for race relations. He taught about racial unity, about the diversity of people, and about their attitudes and tolerance toward other people. He spoke to people about overcoming their fears and prejudices.

A parallel exists between the attitudes Dr. King worked to modify and those some people have toward environmental issues. The parallel is elementary—people discriminate because of ignorance and a lack of familiarity or awareness of another's place in the world. Fear, begot by ignorance, is the primary wrapping on a package known as prejudice. The consequences of opening the package are unfair treatment of others, self-induced anxiety, and loss of harmony in the world.

A variety of groups, both human and wild, have suffered from environmental discrimination, some over the last few years or decades, others for centuries. Among the nonhuman victims have been snakes, wolves, and mountain lions. Less apparent in some instances are particular groups of people who have been victims of prejudice, including environmentalists, timber companies, and private landowners. All these groups, including the animal predators, have individual members who have done no harm. Ill feelings toward them come from the fear created by a few and ignorance about the group as a whole. Every one of these groups has members who have done, and will continue to do, positive things for the environment.

A common feature leading to prejudice and discrimination against a group is a negative action by a single member or faction that is viewed as characteristic of the group as a whole. A principle of human behavior is that we judge others in our own group by their individual actions, whereas we judge other groups by the most conspicuous traits displayed by a few individuals.

If environmentalists blow up a whaling vessel, some people then categorize anyone who protests whale hunting as an extremist. If a mountain lion kills a domestic sheep, some ranchers conclude that all mountain lions should be eliminated. If a private landowner destroys woodlands and wetlands on her property, some people get the mistaken notion that private landowners have no regard for environmental stewardship.

Ignorance and irrational fears about groups of people or other animals lead to negative attitudes toward them and acts of discrimination against them. What is the solution for overcoming this situation? The formula is a simple one: get to know them better.

For example, most snakes are nonvenomous and completely harmless to humans; they play important roles as both predators and prey in natural ecosystems and have lifestyles that can be a source of fascination. Likewise, wolves and mountain lions are natural predators that trim the weak or sick from prey populations. They have captivating patterns of behavior as individuals and only intrude on man's domain where man has already intruded on theirs. But people who are ignorant of the group characteristics of these animals are likely to base their opinion on the behavior of a few individuals.

As for the human groups I mentioned, most environmentalists are not reactionaries who are unwilling to negotiate or compromise about environmental issues. Many timber companies use sustainable forest approaches, are concerned about the entire forest ecosystem, and have individual employees who are as ecologically minded as any research ecologist. And the majority of private landowners are good environmental stewards because they appreciate healthy ecosystems as much as any ardent environmentalist.

Members of these different groups should cultivate an awareness of the ideals and goals of the others. They should recognize and respect the differences that separate them, while searching for common ground. In the spirit of Martin Luther King Jr. we should all champion the diversity we find around us, in our own species and in others, by increasing our knowledge about different groups. Fear, begot by ignorance, can be overcome, because ignorance can be remedied.

Make Your Environmental Resolutions for the Coming Year

So you've made your resolutions for the New Year: lose weight, exercise more, and never again watch another negative political advertisement. How about making resolutions in addition to those perennial favorites? Below is a ten-point checklist of easy-to-make, no-need-to-break environmental resolutions. If you have children or grandchildren or are a teacher, encourage a child to join you in fulfilling the resolutions. By doing so, you might well make a lasting contribution to environmental education.

1. Put this list on your refrigerator, the reminder board in your office, or a bulletin board in your classroom. If you take it to class after the holidays, ask your teacher for extra credit. If you are a teacher, consider awarding extra credit for each item a student completes.

2. Take a walk around your neighborhood or a park with the intention of carefully examining the bark of trees. Compare the color and texture of different kinds. Do they all have the same kind of lichens? (If you aren't sure what lichens are, find out.) Looking closely at some of the basic features of the natural world is educational—and fun.

3. Read a natural history book that discusses a particular group of animals or plants. Many excellent wildlife guides, photograph-filled coffee table books, and even children's stories will qualify. If you are a teacher who's completing this checklist in a classroom, have the students give short accounts about the ecology of a plant or animal mentioned in the book. If you are the parent

of a preschooler at home, read them sections from your favorite natural history books.

4. You may want to wait for warm weather for this one, but it will be worth the wait: pretend you are a behavioral ecologist. Find an animal in your yard and observe it continuously for five minutes. Whatever you pick—insect, squirrel, spider, bird—watch it closely. It just might do something interesting and unexpected. If the animal does not move for five minutes, it doesn't count. Pick another one.

5. Most people are used to enjoying nature through the sense of sight, by looking at plants, animals, and habitats. Learn to observe a habitat with your other senses. Listen for animal sounds, the rustle of tree leaves, or running water. Crush the leaves of trees or shrubs to see

Q: I have a brown skink exploring between my basement and the first floor of my house. My little two-year-old almost had a panic attack this morning as she came running to me about a "MONSTER!" (This was before I encountered the little skink in the basement just a moment ago.) How can I safely remove this little critter from my house? I have been holding my little girl on my hip all day long because she is so afraid of this skink.

A: The easiest way to remove a brown skink is to catch it by hand (the tail will probably break off) and put it outside. It might try to bite, but they are so small it does not hurt when they do. As far as the child goes, I would start showing her pictures of lizards and snakes in a book, explaining that they are just a natural part of the world and have no intention of hurting anyone. Go outside and look at insects, spiders, birds, and other wild things with her. Sometimes children develop fears of certain animals because they have not been taught that most animals mean us no harm. Good luck.

if any have a distinctive scent. Compare the feel of different grasses, leaves, and bark. If you are absolutely certain that a plant is not poisonous, you might even taste it. Learn to use all of your senses when you are in the woods or at a lake.

6. During the colder months you can visit a natural history museum, nature park, zoo, or public aquarium. You probably live within an hour or so of one of these and most of them have an environmental theme of one sort or another. For classrooms, themes such as endangered species, water quality, and overall environmental awareness can be studied in advance, making the visit even more enjoyable.

7. Pick an animal and a plant in your region that you are likely to see on a regular basis (maybe a gray squirrel, crow, dogwood tree, azalea). Read about the natural history of both in at least one source that is not on the Internet, such as an encyclopedia, natural history magazine, or nature book. Then visit an ecology website to learn more about the plant and animal. Remember that most websites have not been subjected to rigorous scientific and editorial reviews. Misinformation is rampant on the Internet, so go to several unquestionably reputable websites and read about both species. Most sites sponsored by a university, museum, or government agency would classify as "reputable." Using the scientific name of a species in the search usually helps in pulling up sites that are more likely to be trustworthy. Learning about the ecology, geographic distribution, and close relatives of a target species will ensure that you appreciate it for the rest of your life.

8. One way to have a long-term impact on the environmental awareness of others is to teach a child to appreciate the natural world. Ask children to think about the plants and animals in the yard or neighborhood. Ask

questions, such as why are so many kinds of birds red, yellow, or blue, but most mammals are a shade of brown, gray, or black. Look up a topic about nature together in a book or encyclopedia. Having a positive influence on the environmental awareness of a child might be the most important action you ever take in behalf of the world's habitats, plants, and animals, not to mention in behalf of ourselves.

9. Help a nonprofit environmental organization. You can make a monetary contribution or donate your time. You might have the greatest impact (and the most fun) by volunteering with a group of friends or classmates to help a local environmental group. Another approach is to provide some tangible support for environmental education at a local school. This is an obvious resolution for science teachers in any grade because that is what they already do. But such a commitment is also appropriate for a schoolchild's parent. You can donate your time to a school ecology program. Taking children on outdoor field trips, even if it's just on the school site, is an area in which most of today's school programs fall short. Ask the science teacher what field trips are planned and if you can help by being a class chaperone. If you do not have time to invest in such an endeavor, offer to buy something for the school. Donating a natural history book or giving a subscription to a nature magazine would help contribute to environmental education.

10. Give this checklist to a friend or relative and ask them to complete it. The more people pay attention to the natural world, the better off the environment will be. When you check off this last item, you can feel good about having kept some meaningful resolutions.

Happy New Year!

Why Does a Pine Tree Produce Turpentine?

Why would a tree living in a habitat that catches fire every few years produce turpentine, a highly flammable substance? That question was asked as I was building a fire at home with a piece of fat lighter, wood from the stump of a long-dead pine tree. Fat lighter, also known as fatwood, catches fire immediately and burns longer and hotter than the driest wood. With a piece no bigger than a cell phone, you can start a fire without paper. Before you strike a match to fat lighter, smell it. Good fat lighter is permeated with turpentine. The turpentine neither harms nor aids the tree while it is alive, becoming of value to it only after the tree dies. Does that sound like a riddle worthy of the Sphinx? The answer to the apparent conundrum lies in the natural world's extraordinary ability to adapt and evolve.

Turpentine, a substance characteristic of pine trees and other conifers, is composed of a mixture of resins and volatile oils. The by-products have been used in a wide variety of applications including caulking for wooden ships, solvent for paint and varnish, and as an ingredient in insecticides, cleaning agents, and shoe polish. Turpentine products have even been used for medicinal purposes. A great turpentine industry was once centered in the South, where pine trees, especially longleaf and slash pine, were tapped for turpentine, the way sugar maples are tapped for sap to produce maple syrup. The turpentine industry took advantage of a pine tree's natural response to injury. If the bark is broken, the tree begins to ooze sticky, yellowish sap that eventually dries and seals the wound with a layer of resin. The material is resistant to most wood-eating insects that might further damage the tree. The liquid can be distilled to produce turpentine.

But longleaf pines also have a characteristic that makes turpentine production seem counterintuitive. They live in what is known as a fire climax community. This means that, historically, trees and other plants that persisted in a longleaf pine community had to survive natural, periodic fires that swept through the forests, primarily as a result of summer lightning strikes. Some ecologists criticize forest management programs that prescribe controlled burns during winter because natural fires would usually have occurred in summer. Presumably, plants and animals in regions that historically experienced frequent fires evolved to tolerate warm weather fires.

Q: We were wondering if the warmer winter we had in Georgia one year could explain the increase in the number of anoles we found around our house and yard the following summer.

A: Warmer winters could possibly result in more insects for anoles to eat in the spring, which could lead to an increase in reproduction in the species and, therefore, more lizards. However, determining the exact cause of an increase or decrease in the population numbers of animals is extremely complex, even for population ecologists who study a particular species in a prescribed area. Part of the problem is the difficulty in determining whether an observed effect (such as change in population size of the lizards) is the direct result of weather or climate changes that affect the animal itself or an indirect result of an effect on another species (such as a parasite or predator) that might influence population size. To further complicate matters, some changes in numbers of an animal species, even over a several-year period, may simply be coincidental with an observed environmental change that is unrelated to the species.

Longleaf pine is a species well-adapted to survive fires at intervals of less than ten years. Young longleaf seedlings, in the so-called grass stage, can be burned back to the ground and then, unharmed, resprout the same season. A larger, more mature tree is also immune to a fast-burning forest fire because its thick bark is resistant to fire (and has no turpentine in it).

But why would a pine tree that, under former natural conditions, was sure to be subjected to numerous fires during its lifetime be saturated with readily flammable turpentine? An ecologically harmonious answer is that the turpentine is advantageous to the tree after it, or any part of it, dies. Here's how. A pine tree dies, and within a few months or years, after the tree's bark has fallen off, a fire sweeps through the area. The dead tree, especially the stump of fat lighter, burns to the ground along with any dead needles or limbs that were already on the ground. Nutrients bound inside the dead tree are returned to the soil and once again become available for other pine trees.

But animals and plants, including pine trees, are not altruistic, so why would this be of advantage to the tree? The simplest explanation is that most of the nearby trees would be descendants of the burned tree. The tree would be returning the nutrients to its own kin. In addition, adding fuel to the periodic fires would eliminate other trees that were not fire-tolerant species and that might otherwise compete with the pine trees.

So there you have it. Pine trees have worked out an efficient and effective mechanism to deal with periodic fires over evolutionary time. The riddle of the turpentine-saturated pine tree is solved, and people benefit if they can find a fat lighter stump to use in their fireplace before the next forest fire comes through the area.

Why Do Animals Turn White in the Arctic but Not the Antarctic?

Why do so many animals turn white in winter or stay white all year at the North Pole but not at the South Pole? Wouldn't the same camouflage conditions apply to both regions?

Throughout the animal kingdom, in circumstances where being unseen or at least inconspicuous is beneficial, camouflage is critical for prey and predators alike. Clearly, both polar regions have plenty of snow to make a white body effective camouflage under certain conditions. However, the Arctic includes vast areas of terrestrial habitats, such as Alaska, Canada, and Russia, whereas Antarctica has no physical connection to another continent. Consequently, during the summer 90 percent of the Arctic is thawed out, whereas more than 90 percent of Antarctica is permanently covered with ice and snow. The ecological consequence is that the Arctic has a variety of strictly terrestrial mammals and Antarctica has none. Therefore, the strategies of native inhabitants in the two regions vary greatly, which means some animals at the North Pole must change their appearance seasonally to be camouflaged.

Penguins, which are found at the South Pole but not the North Pole, are primarily black and white with a dark back and lighter colored front. Being solid white would serve little advantage to an animal that has no land predators to hide from. Likewise, no terrestrial predators are present at the South Pole that need to be white so they can sneak up on unsuspecting prey.

The tactics for avoiding predators and catching prey are entirely different in the Arctic. The Arctic is defined

Q: I live in Tennessee and work in an office that has an attic in which there are brown recluse spiders. The pest control man who sprays the office said the only way to kill a spider with the spray is to get a direct hit. He says the residual spray is ineffective in controlling them. Is this true? The bug man sets out sticky traps for the spiders; the traps do catch the spiders, as well as other bugs.

A: It is probably true that direct spray would be more beneficial in eliminating a brown recluse colony, but the first consideration is whether the spiders are really brown recluses or one of their look-alikes. (This is where a spider identification field guide comes in handy.) The exterminator should also check for egg sacs after the initial spray—if the egg sacs persist, you will have the same problem next year.

as regions of tundra and permanent ice cover above the tree line in North America, Europe, and Asia. Being seen would be a disadvantage for prey animals like Arctic hares (found from northeastern Canada to Greenland) or snowshoe hares (found from Alaska through southern Canada and the northern United States). They fare best in a brown coat during warm months when the soil surface is visible. In the cold Arctic portion of their geographic ranges, when autumn snow begins to whiten the surroundings, white is the best color for a coat. Seasonal color change from brown to white also occurs in some non-Arctic species. For example, the white-tailed jackrabbit is a species that does not occur in the Arctic but is found in northern portions of the western United States and into southern Canada. In the coldest and snowiest parts of their geographic range, they turn white in winter.

The Arctic fox adopts a similar seasonal color change

tactic for two reasons. First, Arctic foxes do not want to be seen by rodents and hares, which are the prey of these wily predators. Second, Arctic foxes often scavenge in winter, dining on the remains of seal or fish left by polar bears. Presumably, polar bears would eat a fox as well, so having a wintertime camouflage of pure white (or sometimes bluish-gray) fur works well for the fox. In summer, brown is in vogue, as with the hares and many other Arctic mammals.

The white coats of Arctic animals have long been popular in the fashion industry, and a magnificent white ermine coat was once the most elegant of winter apparel. A white ermine, with only a black tail tip, must be caught during the Arctic winter. In the summer the ermine turns mahogany brown. Members of the weasel family, ermine are best known among wildlife biologists as short-tailed weasels. For some reason, I have never heard of anyone boasting about the sumptuous weasel coat she wore to the ball.

Valentine's Day Signals the End of Winter Dormancy

"In the spring a young man's fancy lightly turns to thoughts of love." So said Alfred, Lord Tennyson. And though he may not have realized it, as spring approaches, courtship behaviors are prevalent throughout the animal kingdom.

One of the most vivid displays of Valentine red is that of the blue-tailed skinks of the eastern United States. The largest of these is the broad-headed skink. The commonly seen juveniles have a metallic blue tail and bright yellow stripes, a color pattern that remains in a more subdued fashion in the adult females. The males, which become enormous by typical southeastern lizard standards, develop shiny, coppery brown bodies. In the springtime, not too long after Valentine's Day in the southern parts of their range, the male broad-headed skinks begin their courtship. At this time their head and neck turn brilliant red, making them look rather like a Valentine heart moving through the forest. Is it coincidental that so many animals begin courtship around Valentine's Day or is there some significance to the seasonal timing?

A connection between Valentine's Day and spring mating rituals of some animals sounds like a romantic idea, but no valid relationship exists between our human celebration and the reproductive patterns among wild species. The origins of Valentine's Day celebrations, which now occur virtually worldwide, are disputed and almost certainly based on tales that qualify more as myth or fable than as documented events. For human beings, Valentine's Day is more or less an artificially created reason to have a celebration, and the idea has spread throughout the world.

Q: Having recently moved to Louisiana, I am dismayed by the many different kinds of spiders I have seen. Are they all dangerous? Should I try to get rid of any I see around my house?

A: More than forty thousand different kinds of spiders have been described worldwide, so you will find them almost anywhere if you look hard enough. Instead of being concerned about them, embark on a voyage of discovery. Start by learning about some of the major groups you are likely to find in your area, such as orb weavers, or crab, wolf, and jumping spiders. The more you learn about spiders (and their webs), the more you will appreciate them. They are among nature's most awesome creatures.

February 14, the date Americans have selected for a celebration of Cupid's matchmaking antics, happens to occur at a time when native wildlife species in the north temperate zone begin to reemerge after winter dormancy. The days have been getting longer since the darkest days leading to December 21 and will continue to do so until June 21. More new-growth vegetation means food is available for herbivores of all sorts, from insects to mammals. And more food means more energy that can be devoted by females to production of young and by males for territorial pursuit of females. Courtship and mating occur throughout the year for some plants and animals, but the burgeoning resources associated with the advent of spring assures major activity. Valentine's Day is well-timed for the phenomenon.

Among many birds—from Florida northward, depending on the latitude—evidence of courtship can be seen around Valentine's Day. Male bluebirds begin to show a hint of the impressive plumage soon to come, and male goldfinches start changing from drab olive to stunning yellow. The females of these and many other species of birds remain comparatively drab, while the males use their plumage to attract females. Male displays of breeding colors in some species might be considered the equivalent of brightly colored Valentine cards intended to appeal to the object of your affections.

Turtles do not get left out during Cupid's season either. The mating period for most temperate zone species begins in late winter or early spring as temperatures begin to rise and males begin seeking females and engaging in courtship activities. Turtle biologists assume that all species have a ritualized courtship process that leads to the mating event itself, but courting behavior in the wild has been observed in only a small proportion of the world's turtles. Nonetheless, it assuredly happens.

Turtle biologists have observed complex and intriguing courtship rituals in some species. Adult male painted turtles and slider turtles (including the common red-eared sliders often kept as pets) have elongated foreclaws used in an elaborate courtship behavior called "titillation." During this remarkable ritual, the male extends his front feet and turns them so that the backs are touching. Then he vibrates his long claws in the water in front of the female. A female interested in mating will follow the male, who slowly swims backward. One of the best opportunities to observe the fascinating process of titillation is at a public aquarium in which freshwater turtles can be seen through glass in natural underwater settings.

For almost any form of human behavior that can be identified, an equivalent or near-equivalent can be found somewhere in the animal kingdom. Spring mating rituals among many animals prove that courting is no exception.

Conclusion

Survey after survey indicates that most people favor protecting our natural environment. People want clean air and clean water, and they want to know that our native plants and animals will be around for future generations. They also acknowledge that thwarting selfish attempts by a few individuals to diminish environmental protection is a worthy pursuit.

But not everyone agrees on how to achieve such goals. And, as we point out in the introduction, some of the eleven classic attitudes humans have toward wildlife are in direct conflict with each other. Someone with a naturalistic mindset is unlikely to find common ground with a person who is indifferent or negative toward all nature. But such polar opposites are the exception. Most of the attitudes have room for compromise.

We have long asserted that head-to-head combat on environmental issues does not, in the main, serve our native plants, animals, and habitats well. Unfortunately, compromise has become, for some people, a sign of weakness. We are not sure exactly where or when the idea of compromise became something to shun, but we believe that our natural world will be better served by compromise than by intractability. Compromise does not require forsaking one's principles or selling out. It does mean sometimes bending or refocusing one's own views and positions to better understand those of others.

We are surrounded by countless life forms that are educational, thought-provoking, breathtaking, inspirational, beautiful, or bizarre. The more people realize just how much the natural world has to offer, the more likely we are to make environmental concerns a top priority. So the next time you have the opportunity to go for a hike in the woods, take a kayak down a stream, or just walk outside in your own neighborhood, take it. Nature is open 24/7 every day of the year. Enjoy!

Index

Bigfoot, 114
biodiversity hypothesis, 125–26
birds: and antennas, 121; color patterns of, 51; and crocodilians, 59, 62; eggs of, 29; falling from nest, 25–27; feeding of, 90; and mating, 28; and mistletoe, 154–56; and siblicide, 14–17; transporting plant seeds, 154. *See also names of specific birds*
blackbirds, 159
blackbirds, redwing, 51
black walnut trees, 74
black widow spiders, 140
Blanding's turtles, 48
bluebirds, 185
bobcats, 1, 53, 104–6
bobwhite quails, 15, 74
bower birds, 147
box turtles, 48, 61
brown recluse spiders, 182
buffalo, 165
buffalo, African, 115–16
Buhlmann, Kurt, 101, 103
bullfrogs, 57, 101–3
bumblebees, 18–22, 24

calling birds, 157
camouflage, 1, 51, 181
Canada geese, 28, 128, 158
canebrake rattlesnakes, 29, 39, 135, 137
cardinal flowers, 101, 103
cardinals, 12
caribou, 89, 164–67
carotenoid, 128–29
carpenter bees, 22–24
Carr, Ann Bartram, 160
Carr, Robert, 160
catamounts. *See* mountain lions
caterpillars, 76
cats, 87
cattle, 111–12, 165
cedar waxwings, 54, 155
Chelsea Historic Site (MD), 151
chipmunks, 90, 168
chlorophyll, 127–29
chuck-will's-widows, 70

goats, 165
golden silk spiders, 109
goldfinches, 51, 186
Grand Pacific Glacier, 83
grizzly bears, 89, 167
groundhogs, 168–70
gulls, glaucous-winged, 84
gulls, herring, 96

habitat preservation, 117–19, 122
hairworms, 108–10
hares, 183
hares, Arctic, 182
hares, snowshoe, 52, 182
Harris, Keith, 137
Harris, Nick, 69–71, 137
Hemingway, Ernest, 115
hemlock forests, 84
herons, 79
hippopotamuses, 116
holly trees, 151–54
honey, 20
honeybees, 18, 20, 22
hornets, 24, 131–33
horsehair worms, 108–10
horses, 111–12, 165
house finches, 12
hummingbirds, 103
hunters, 117–19, 139
hyenas, 116

Ilex aquifolium (European holly), 152
Ilex opaca (American holly), 151
insects, 33, 54, 152, 154. *See specific names of insects*
invasive animals, 120
invasive plants, 120
Invasive Pythons in the United States (Dorcas and Willson), 79
Irving, William, 156

jackrabbits, white-tailed, 182
jaguars, 104, 106
Jamaican frogs, 11–12
jellyfish, 76